Debra Carbarnes

P9-CAT-800

STINK ?
— NATURE CENTER —

Fall

Winter

A Guide to
Observing Insect Lives

A Guide to
Observing Insect Lives

DONALD W. STOKES

ILLUSTRATIONS BY DEBORAH PRINCE

LITTLE, BROWN AND COMPANY — BOSTON   TORONTO

FIRST EDITION

LIBRARY OF CONGRESS CATALOGING IN PUBLICATION DATA

Stokes, Donald W.
    A guide to observing insect lives.

    Bibliography: p. 367
    1. Insects—Behavior.   2. Insects—Seasonal
distribution.   3. Wildlife watching.   I. Title.
QL496.S86 1982   595.7′051   82-23927
ISBN 0-316-81724-4

MV

Designed by Susan Windheim

*Published simultaneously in Canada
by Little, Brown & Company (Canada) Limited*

PRINTED IN THE UNITED STATES OF AMERICA

To all who enjoy a closer look at nature

Contents

Fall Insects 195

Winter Insects 279

A Guide to
Observing Insect Lives

Watching Insects

I used to assume that the only thing insects could ever do for me was to create an uneasy feeling up and down my spine. I certainly never looked for them, and when I happened to find one I usually avoided or swatted it. But as a naturalist, I am repeatedly discovering that all types of life hold some fascination for the careful observer, so at one point I decided to give insects a fair chance. I started to read about them, learned how to distinguish the basic types, and then began to observe their habits. As I continued to learn through observing and reading, I became less hostile and more fascinated; basically, my curiosity overcame my initial uneasiness. Now, a few years later, I actually seek out insects in all seasons and find watching their behavior a superb activity, an experience that has provided me with some of the most absorbing moments in my life as a naturalist.

A *Guide to Observing Insect Lives* will help you experience some of these moments for yourself. Unlike a traditional field guide, which instructs you to find an insect and bring it back to the book for identification, this book leads *you* to the *insect* and then centers and focuses your observations on the insect's most intriguing behavior.

Insects are exciting to observe, for they are among the most accessible of all animals. They are common in just about every environment, they are easy to find, they generally do not move very far from one spot, and they usually allow you to get quite close. Besides this, their behavior is often seemingly bizarre and unusual and in striking contrast to our own. Some people find this unnerving; I find it refreshing. It reminds me that other animals, although in some ways like us, are in most ways amazingly different.

How to Use This Book

This book is divided into four seasonal sections. In each section there is an introduction to watching insects, a location guide, and descriptions of the lives and behavior of specific insects. At the beginning of each insect entry, two important illustrations show the typical habitat in which you will find the insect, and a life-size picture of the insect. Along with these, there are four brief paragraphs, describing the relationship of the insect to other insects, the insect's life cycle, the highlights of its life cycle, and how you can find the insect during the highlight of its life cycle. Following this is an extended section on what you can observe once you have found the insect.

Although this book can be just enjoyable reading, it is designed to help you find our common insects in the field and observe their behavior. This chapter and the next will introduce you to watching insects and to the basics of insects' lives.

The best way to begin your study is to choose an insect in the proper season, look at its habitat and life-size illustrations, and read the section on how to find it. The location guides at the beginning of each seasonal section are quick references to help you know at a glance which insects you are likely to find in a particular environment. By reading the introduction to the season and the other material on the insect you want to study, you will increase your chances of success in finding it.

Once you have found the insect, read the section on

what you can observe, to help you to see the insect's most interesting features.

Finding Insects

If you are just looking for insects in general, one method of finding them is to go to an area with a variety of plants, such as a field, woods, or garden, and sit in one spot for five to ten minutes. Look closely at the plants right around you and soon you are likely to see insects crawling on leaves, flying about between plants, or even landing on you.

You can also search more actively for them. The best place to do this is in lush, weedy areas. Look under leaves, around flowers, along stems, and around the base of the plants. Remember, insects are small. To see and appreciate them, it is best to get very close to them, at least within a foot, unless you are observing the social bees (honeybees and bumblebees) and social wasps (paper wasps, yellow jackets, and hornets), which can sting. It's wise to keep your distance from them, especially when you are near their homes.

There are several tools that will help you get a better look at an insect. The one I find most useful is an old, clear glass, wide-mouthed jar with a lid, such as a peanut butter jar. You can trap many of even the fastest fliers in such a jar and have a chance to watch them more closely. Sometimes a twig or weed stem in the jar enables the insect to hold on to something and remain still.

Two other tools are a net for catching fast fliers or scooping up water insects, and a magnifying glass to get a look at the details of insects' coloring and structure. In all of your encounters with insects, be sure to return them to their natural habitat as soon as you are finished observing them. Take care with them: they are an integral part of the environment, just like you.

Observing Behavior

Many people see insects, but few ever observe their behavior. To observe the behavior of an insect, all you have to do is watch it for a least two or three minutes. At first, this may seem like a long time to watch any one thing in nature, especially if you are used to taking walks and not stopping, or used to just identifying and then going on. But I assure you that in the long run this small investment of time will repay you again and again with wondrous glimpses into insects' lives.

There are many types of insect behavior to see. One of the most exciting is social behavior — interactions between individuals of the same species. Social behavior usually involves some form of communication. Two good examples of this are the signal system of flashes used by fireflies to attract mates, and the marvelous activities of dragonflies as they guard territories and choose mates. The hunting behavior of insects is equally engaging. Some insects are active hunters, such as the robber fly, which darts out from perches to catch other insects in midair. Others are passive hunters, such as the ambush bug, which waits, camouflaged, among goldenrod blossoms and catches insects that come to the flowers for food. All insects eat and many do it elaborately, leaving behind a great deal of evidence as to how and what they ate. For example, the pine-tube moth ties pine needles together into a tube, and then bites them off at half their length, pulls them into the tube, and eats them. Acorn weevils burrow into young acorns and lay their eggs in them. The young feed in the developing acorn and then burrow out after it has fallen, leaving just the empty acorn shell with a hole in it.

Other insect behavior involves building or altering environments. You can observe the web nests of the tent caterpillar, the sand pits of the antlions, and the beautifully detailed cases of the bagworm. Still other insects have unusual ways of caring for their young, such as the ambrosia beetle, which makes tunnels in wood and raises fungus to feed to its young; or the twig girdler beetle, which lays eggs at the tip of a branch and then girdles the branch so that it breaks off and falls to the ground. The wood begins to decay and becomes just the right consistency for the young to eat.

These examples are not just for you to read about and marvel at from your fireside chair; they are things you can easily see for yourself. You do not need any fancy equipment, a wilderness environment, or special knowledge. All you need is the help of this guide and that spark of energy that will get you outside and observing.

What Is Known

A quick look at this guide may suggest that a lot is known about our common insects, but as you start to observe, you will see that the gaps in knowledge are great and that the questions far outnumber the answers. This is because insect behavior is still a relatively unexplored frontier for scientists and naturalists, and there is still so much to learn. After observing insects for a while, I am sure you will think of many good questions that have not yet been answered. Don't be discouraged at not being able to find the answers in other books or articles; just go out and observe for yourself. After all, most of the information already in this book comes from people just like you — curious, intelligent, and willing to stop along the side of a road or path for those few extra moments that reward them with an ever-increasing knowledge and understanding of insects' amazing lives.

The Basic Facts about Insects

Before starting to observe insects, there are a few facts about them that will be immensely helpful to you. First, you need to be able to recognize an insect when you see one; second, you should know the two basic kinds of insect development; and third, you should have some familiarity with the eight most common orders of insects, for they contain most of the insects you will find. This information is really surprisingly easy to master and will give you a strong framework on which to place your observations.

How to Recognize an Insect

Since there are a variety of other small animals with which insects might at first be confused, such as daddy longlegs, fairy shrimp, millipedes, and spiders, the first question in identification is: How do I distinguish an insect from all other animals? Luckily, adult insects have two distinguishing characteristics:

1. Adult insects have three major divisions to their bodies, called the head, thorax, and abdomen.
2. Adult insects have three pairs of legs, all attached to the

thorax, or second body division. (Spiders have only two body divisions and eight legs, and so are not insects.)

Many adult insects have one or two pairs of wings, and these are also attached to the thorax.

On the other hand, the immature stages of insects are incredibly varied and they have few obvious features in common. In some cases you may see three body parts and three pairs of legs, but in others there may be no legs or many pairs of legs and the body parts may be obscured. In these latter cases it may be difficult to be sure you have found an insect.

Insect Development

Insects change a great deal during their lives, in many cases so much that the various stages of their development look entirely different. These changes are called metamorphosis: meta = change, morphe = form. The two most common types of metamorphosis are called complete and gradual.

In complete metamorphosis, when the insect hatches from the egg it is called a larva. Larvae shed their skins from time to time as they outgrow them and usually resem-

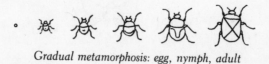

Complete metamorphosis: egg, larva, pupa, adult

Gradual metamorphosis: egg, nymph, adult

ble small worms throughout this stage. After a certain number of sheddings, their outer skin hardens into a tough casing and the insect is now called a pupa. (Some insects construct a covering over their pupa and this is called a cocoon.) During the pupal stage (pupation), the insect undergoes a radical change in form. When it has finished these changes, it emerges from the pupal case and is now an adult and able to reproduce.

In gradual metamorphosis, when the insect emerges from the egg it is called a nymph. Nymphs also shed their skins as they outgrow them, but instead of looking like small worms they usually look much more like an adult, except that they lack fully developed wings. With each shedding they look more like the adults, until finally they have functional wings, have stopped all growth, and are able to reproduce. At this point they are called adults.

The Eight Major Orders of Insects

There are twenty-six orders of insects, but the vast majority that you will find belong to only eight of these. By knowing these eight orders — how to identify them and a little about their life habits — you will find that you know more about insects than most nature enthusiasts, and even most naturalists! It is amazing how accessible this information is and yet how little it has become a part of the public's knowledge. Below is a short guide to the identification of the eight major orders.

HYMENOPTERA ("MEMBRANE-WINGED"): BEES, WASPS, ANTS

All of these insects have two pairs of thin, clear, membranous wings, and in the females, the abdomen ends in a well-developed egg-laying organ and/or stinger.

From left to right: *wasp, bee, ant*

You may wonder why ants are in this group. They also have wings but only at certain stages of their lives; many have stingers as well. Some flies look like bees or wasps, so check for two pairs of wings before assuming that an insect belongs to this group.

Metamorphosis is complete.

DIPTERA ("TWO WINGS"): FLIES, MOSQUITOES, GNATS

If you find an insect with just one pair of wings, then it is a fly of some kind. Not all flies look like houseflies; some

are delicate, such as mosquitoes, some look like bees or wasps, and some are small, such as gnats.

Metamorphosis is complete.

LEPIDOPTERA ("SCALY WINGS"): MOTHS, BUTTERFLIES

These insects are easy to know, for their two pairs of wings are covered with small scales that rub off easily.

Metamorphosis is complete.

COLEOPTERA ("SHEATH WINGS"): BEETLES

The main clue to beetles is a pair of hardened wings covering the top of the body and meeting in a straight line down the back. Beetles have two pairs of wings, but you do

not often see both pairs since, when at rest, the membranous hind wings are folded under the tough front ones.

Metamorphosis is complete.

ORTHOPTERA ("STRAIGHT WINGS"): CRICKETS, GRASSHOPPERS, LOCUSTS

It is not hard to identify this group of insects since they usually have long back legs, hop high in the air, and often

make rhythmic sounds. This group also includes the praying mantis and the cockroach. What they all have in common are thin, leathery forewings that cover larger hind wings that are folded like a fan when at rest.

Metamorphosis is gradual.

ODONATA ("TOOTH"): DRAGONFLIES, DAMSELFLIES

These insects have two pairs of long, narrow, membranous

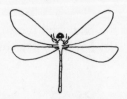

wings that are roughly equal in size. They also have large eyes and extremely long, narrow abdomens.

Metamorphosis is gradual.

HOMOPTERA ("SAME WINGS"): APHIDS, CICADAS, LEAFHOPPERS

These insects have two pairs of membranous wings that are held in a tentlike or rooftop position over the body when at rest.

Metamorphosis is gradual.

HEMIPTERA ("HALF-WINGS"): BUGS, BACKSWIMMERS, WATER STRIDERS

You can identify these insects by a triangle on the back just behind the head. This is formed by the way the insects fold their forewings when at rest. Hemiptera have two pairs of wings: the hind wings are all membranous, while the basal half of the forewings is hardened. Some bugs look similar to beetles, but you can distinguish them by seeing that the

wings form a triangle on the back instead of meeting in a straight line.

Metamorphosis is gradual.

Spring Insects

Observing Insects in Spring

From the beginning to the end of spring, there is a gradual emergence of more and more insects from their winter homes. In general, each species emerges as its source of food becomes available. This is great for the insect observer, for, instead of being swamped by a multitude of species and behaviors, he or she is more at leisure to locate and enjoy the insects a few at a time.

One of the first sources of food in spring is the flowers of willow shrubs, such as the matured catkins on pussy willows. These offer a great deal of pollen and nectar and are good places to find early spring insects, especially solitary bees, bumblebees, and mourning cloak butterflies, although the latter may emerge even earlier and feed on sap flowing from tree wounds. Following the flowers, a few leaves start to emerge on shrubs and trees, and they are immediately fed on by insects. Leaf beetles attack the leaves of willows, and tent caterpillars start to feed on the leaves of chokecherry and black cherry. When the grasses start to get green and a few wildflowers start to bloom, go out into meadows and look for the little accumulations of spittle on plant stems; inside these are the spittlebug nymphs. Also in the meadows there may be the first white butterflies and sulphur butterflies feeding on the flowers and engaging in mating displays. And a little later, check the nearest ponds or streams for water striders and whirligig beetles on the surface of the water. They can now become active because their food — small flies and other insects on the water surface — is abundant.

By mid-spring several insects will become active right

around your house. Two will be found on your windows or screens at night. One will crash into the windows and make a lot of noise as it crawls about; this is the June beetle. The other is quiet and you will have to look for it on your screens; it is an ichneumon wasp. During the day, two other spring insects will be active around the house. One is the paper wasp, gathering bits of wood with which it makes its brood cells, often under house eaves. The other is the antlion, making pits in dry, sandy areas and trapping insects in them. By the end of spring, hundreds of insects are emerging every day and it becomes harder to choose any one species to observe, for the air is full of marvelous distractions.

SPRING INSECT LOCATION GUIDE

IF YOU ARE NEAR

LOOK FOR

	FIELDS	PONDS & STREAMS	WOODS	FIELD EDGES	HOUSES	BARE GROUND
ANTLIONS					■	■
SPITTLEBUGS	■			■		
WATER STRIDERS		■				
WHIRLIGIG BEETLES		■				
JUNE BEETLES			■		■	
WILLOW AND COTTONWOOD LEAF BEETLES		■				
WHITE AND SULPHUR BUTTERFLIES	■					
MOURNING CLOAK AND TORTOISE SHELLS			■			
TENT CATERPILLAR				■		
ICHNEUMON WASPS				■	■	
ANTS	■				■	■
PAPER WASPS		■			■	
SOLITARY BEES				■		■
BUMBLEBEES	■		■			

ANTLIONS

Relationships

Antlions are a family (Myrmeleontidae) of insects in the order Neuroptera, or nerve-winged insects. All members of this order go through the stages of complete metamorphosis, that is, egg, larva, pupa, and adult; and as adults they all have two pairs of transparent wings with many fine veins all over them. Another common insect in this order is the lacewing (see Aphid, summer section).

Life Cycle

Antlions overwinter underground as larvae. In spring they create pits in the sand in which they catch prey. The larva molts three times, and it may take from one to three years to complete the stage. Each summer of the larva stage, the larva creates a pit for catching prey. When mature, it makes a cocoon at the base of its pit. This occurs in spring or

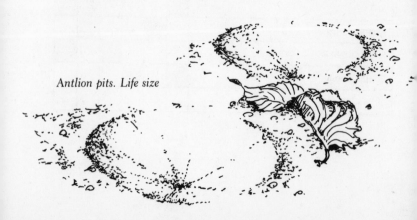

Antlion pits. Life size

Antlion habitat

summer and pupation lasts about one month. The adult emerges in summer, feeds and mates, and the female lays eggs singly in sandy areas. The eggs hatch and the tiny larva starts to make a pit in which it will get its first meal. When winter comes, the larva burrows slightly deeper and remains there until spring.

Highlights of the Life Cycle

Undoubtedly the most enjoyable time to observe antlions is when they are larvae and live at the base of their little pits. There are all kinds of things you can do to see their behavior, including scooping up the insect for a closer look. The adult phase can be seen but usually only by chance.

How to Find Antlions

Antlions are easiest to find in their larval stage. You don't look for the insect; rather, you look for the little pit that it makes. These are found in dry, fine-grained soil, usually in spots protected from rain, such as under house eaves, bridges, or rock ledges. The pits are circular, up to an inch deep, and from a half inch to two inches across. There are usually many in the same spot. The adults look like damselflies except that they have long antennae and are weak fliers. In midsummer you may find them in the area of the pits, or see them flying around lights at night.

What You Can Observe

BEHAVIOR IN THE PIT

To understand what you will see at the pits, you need a little background on how the pits were made and on what happens during a successful capture of prey.

Right after hatching from an egg, the larva starts to make a pit. It does this by backing around in a spiral and repeatedly flicking soil away from the center of the circle. It finally ends up in the center of the pit. It waits, buried just beneath the sand, at the base of the pit. When potential prey, such as an ant or caterpillar, walks into the pit, it gets stuck, since the angle of the pit walls is such that they

Larva and dug-up cocoon in pit

give way as it tries to escape. The larva makes escape even harder by flicking sand up on top of the prey.

The purpose of this is to get the prey down to the base of the pit where the larva will grab it with pincers and pull it down into the sand. Its front pincers are also tubes through which the antlion injects a paralyzing fluid into the prey, and then sucks out its juices. When the larva is finished feeding, it tosses the empty external skeleton of its prey out of the pit.

There are a couple of ways to see all of this in action. One is to wait for an insect to crawl into a pit. I do not recommend this unless you have a great deal of patience, at least as much as the antlion larva, and a couple of hours at your disposal. Another way is to gently coax the grand scheme of things and drop an ant into a pit. A third way, which is perhaps less bloodthirsty, is to take a small blade of grass or a pine needle and make the tip of it simulate the actions of an insect caught in the pit. You will see the sand on the pit walls keep tumbling down to the center and you may see the larva flick sand up at where you are touching the pit. The antlion may even momentarily grab the end of the pine needle.

THE LARVA

To get a look at the larva, take a scoop of earth that includes the base of the pit. Somewhere within the earth you have collected, you will find the larva. It is from one-eighth to one-half inch long and has large pincers on its head. Its body is covered with tiny hairs, and those on its abdomen angle forward so that once the larva has caught its prey, it can back farther into the soil and it is hard for its prey to pull the other way. This gives the larva a mechanical advantage over large prey. The direction of these hairs makes it difficult for the larva to move forward, and, in fact, some species only walk backward.

THE ADULT

The adult insect is found mostly by chance. It looks a lot like a damselfly except that its antennae are about a quarter inch long (damselfly antennae are hardly visible) and no-

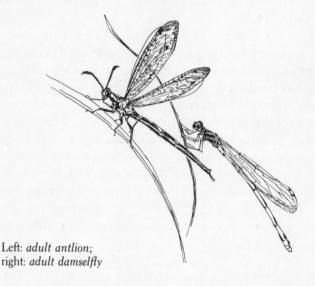

Left: *adult antlion*;
right: *adult damselfly*

ticeably clubbed at their tip. Also it is a feeble flier, unlike the damselfly, which is very agile. Little is known about the adult's life, except that it is often attracted to lights at night, so anything that you discover about it will be valuable. To increase your chances of finding the adult, look in the area of the pits in early to midsummer.

SPITTLEBUGS

Relationships

Spittlebugs are a family (Cercopidae) of insects in the order Homoptera, or same-wing bugs. They are best known for their habit of creating masses of spittle. There are many species and they are difficult to distinguish. The adults look like leafhoppers except that they are stubbier and lack the rows of spines that adult leafhoppers have on their back legs.

Spittlebug habitat

Life Cycle

Spittlebugs overwinter as eggs laid on plant stems. In spring the eggs hatch and the nymphs immediately create spittle masses from excess plant juices they feed on. They undergo several molts within the spittle and may move and form new masses of spittle after some of the molts. After the last molt, they are winged adults and feed in the same manner but no longer create or live in spittle. There are one to three generations per year, depending on the species and the latitude in which the insect lives. Eggs of the last generation, which are laid in fall, overwinter.

Highlights of the Life Cycle

The most obvious stage of this insect's life is the nymphal stage, when it creates spittle. Its ability to create spittle is remarkable and fascinating to observe. The adult can also be seen and observed in just about any large area of tall grasses.

Spittlebug spittle. Life size

How to Find Spittlebugs

You can find spittlebugs in the adult or nymphal stages, but they are easiest to recognize in the nymphal stage, for it is then that they create spittle. Look for the spittle on the stems of plants in lush, weedy areas or on the new, green twigs of pines and other trees and shrubs. By gently pushing the spittle aside, you can find the bug inside; it is usually light green and about one-eighth inch long. The adults are best found by bending down among lush grasses and waiting for the insects to jump on you. They make large hops and are stout, oval insects, usually brown, and about a quarter inch long. They may be induced to come out of the grasses if you brush your hand through the vegetation.

What You Can Observe

THE NYMPHS AND THE SPITTLE

If you examine a number of collections of spittle and look closely at the nymphs, you will see that they are in various stages of maturity. Their wings become more and more obvious the older the nymph becomes. Sometimes several nymphs live in the same mass of spittle.

If you clear away the froth from a nymph and wait patiently, you may have a chance to see the nymph create more spittle. It stands head down on a plant stem, inserts its beaklike mouth into the plant, and sucks up juices to feed on. Excess juices are excreted out its anus in a mucilaginous mixture that falls down over the nymph and is frothed up as it passes over little projections on each side of the nymph.

Creating spittle is a distinctive characteristic of the nymphs of this insect family. The froth may serve a number of functions. It certainly provides a moist environment that

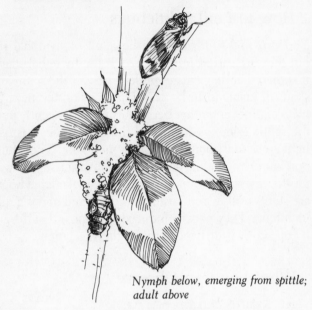

Nymph below, emerging from spittle; adult above

keeps the vulnerable nymph from drying out. It may also provide protection from predators, not by camouflaging the insect, for the spittle is white and obvious, but by providing a distasteful or unpleasant substance that a predator would have to search through. There is, however, at least one species of vinegar fly (*Drosophila sp.*) that spends its larval stage in the froth but does not harm the nymph.

FOUR COMMON SPECIES

There are four common species of spittlebugs, two of which live in meadows and two of which live in pines. You will probably see the spittle from these insects in meadows and on pines.

Two common species in meadows are the meadow spittlebug (*Philaenus spumarius*), whose nymph feeds primarily on clover and alfalfa, and the diamond-backed spittlebug (*Lepyronia quadrangularis*), whose nymph feeds mostly on grasses. The adult of the former is brownish and spotted, the adult of the latter is brown with two darker oblique bands going across its back.

Two common species that feed on pines are the Saratoga spittlebug (*Aphrophora saratogensis*) and the pine spittlebug (*Aphrophora parallela*). The Saratoga spittlebug feeds on pine in the adult stage and then lays its eggs in the dead wood or twigs of the trees. The next spring when the nymphs hatch, they move to the ground and feed on shrubs, especially sweet fern, brambles, and young willows or aspens. When matured into adults, they return to the conifers to feed and to lay their eggs. The pine spittlebug spends its entire life on the trees and occasionally is found on fir and hemlock as well.

WATER STRIDERS

Relationships

Water striders are a family (Gerridae) of insects in the order Hemiptera, or half-wing bugs. They are the largest, most active, and most common bugs that live on the water surface. Two common genera are *Gerris* and *Metrobates*, the former having a long thin body a half inch long or longer and living in mostly still water; the latter having a short, stout body one-quarter inch long and living on the surface of fast-moving streams.

Water strider creating ripples. Life size

Water strider habitat

Life Cycle

Water striders overwinter as adults under rocks or logs at the bottom of streams or ponds. They come to the surface in spring, feed on other insects caught in the surface film, and mate. The fertilized females attach their eggs in parallel rows to the surface of submerged rocks or logs. Nymphs hatch from the eggs in about two weeks and swim to the surface, where they must break through the surface tension and move on top of it. They undergo five molts as nymphs, each taking about a week, before finally becoming adults. There may be as many as three broods per year. The adults maturing from the last brood go underwater to overwinter.

Highlights of the Life Cycle

All aspects of the lives of these insects except the egg stage can be easily observed. Their feeding behavior and mating behavior are among the most interesting features to watch, but you can also enjoy their manner of movement and the physics of their support on the water surface. You can also see the various sizes of nymphs as they develop into adults.

How to Find Water Striders

Just go to the edge of any still area of water, be it a pond, lake edge, or stream edge, and look for insects on the water surface. They are about a half inch long and are supported usually by their long second and third pair of legs. They are thin insects and not immediately apparent when standing still, but obvious when moving.

What You Can Observe

SUPPORT ON THE WATER SURFACE

Most of us have looked at water striders at some time, most likely when we were young and staring absently at a stream

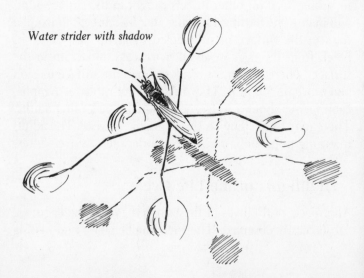

Water strider with shadow

or pond edge. We have watched their effortless gliding over the surface of the water, supported on legs as thin as hairs. One feature that most people notice during their first encounter is that in the shadow of the insect on the stream bottom, there seem to be pads at the tip of each leg. These pads are actually just indentations in the water surface caused by the weight of the insect and the resistance of its feet to breaking the surface tension.

We are so large that the surface tension of water rarely affects us, but the strider's whole life depends on it. Striders are able to stay on the surface because the tips of their legs are lined with many tiny hairs that repel the water, and the claw, typically at the end of an insect's foot, is farther back on the strider's leg so that it does not break the surface tension. Sometimes strider feet do get wet, in which case the bug must crawl out onto vegetation until they dry before the insect can support itself on the water surface again.

MOVEMENT AND HUNTING

Striders look as if they have only four legs, but of course all insects have six; the strider's other pair is short and drawn up beneath its head. Sometimes this pair touches the water and sometimes not. The four long legs are used for locomotion, while the front legs are used for capturing and holding prey that the insect feeds on.

The strider has two types of movement: slow, gliding motion created by rowing the middle legs while the hind legs trail, and quick jumps that involve both pairs of back legs. The gliding motion is used as the strider orients itself toward other insects on the water surface. The quick jump is used as the strider pounces on its prey and grabs it with its smaller front legs.

Prey is detected on the surface through the tiny ripples its motion generates. The striders have sensory organs in their legs that pick up the vibrations and help orient the strider to the prey. The strider rushes over to the prey, grabs it in its two front legs, and jabs its beak into it. The juices pumped into the prey stun it and start to dissolve its innards. When completely dissolved, the innards are sucked out by the strider, and then the remaining skin of the insect prey is cast aside.

MATING BEHAVIOR

From spring to early fall, you will see various interactions between striders as mating and egg-laying occur. The descriptions of mating that follow are from research on non-native species, since native species have yet to be studied.

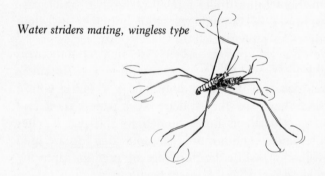

Water striders mating, wingless type

But if you look for yourself, you will see many of the patterns in our native striders. Males start by locating a fixed object, such as a bit of wood or plant in the water, that is suitable for egg-laying. They remain at this spot and give off ripple patterns by moving their middle legs rapidly up and down. These are fine ripples and occur at a frequency of about ten to thirty per second. It is believed that two patterns are used: one that attracts receptive females and one that may warn other males that if the signaler is approached too closely, he will attack. Males seem to defend these sites against other males. When a receptive female approaches, she touches the male with either her leg or mouth and then holds on to the egg-laying object while the male mounts her back for mating. After mating, the female will lay eggs on the spot while the male remains nearby, keeping other males away through ripple signals or direct aggression.

WINGED OR WINGLESS

In most species of striders there are two forms: winged and wingless. The winged forms disperse to new areas of water and are usually found as lone individuals. The wingless forms are usually in groups. Some wingless species live in streams that dry up in midsummer. These insects are able to burrow underneath the mud during the dry season, then reemerge when water is back in the stream.

If you watch the insects closely, you may see some with tiny red dots on them. These are minute red water mites, which are parasites on the striders.

WHIRLIGIG BEETLES

Relationships

Whirligigs are a family (Gyrinidae) of insects in the order Coleoptera, or beetles. They are the only beetles that swim within the surface film of the water, and their whirling about makes them visible from quite a distance. There are two genera, *Dineutas* and *Gyrinus*, the former about a half inch long and the latter more often a quarter inch or less in length.

Whirligig habitat

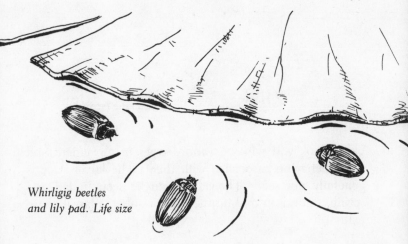

*Whirligig beetles
and lily pad. Life size*

Life Cycle

Whirligig beetles overwinter as adults in mud or debris at
the bottom of ponds. In early spring the adults become
active on the water surface and mate. The females lay eggs
in rows or masses on the stems of submerged plants and
then die. Eggs hatch in about two weeks and the larvae
feed on other insects and small animals underwater for two
to three months. When mature, they leave the water and
make small pupal cells on nearby plants or mud. The pupal
stage lasts about a week, and then the new adults emerge
in mid- to late summer. They form large aggregations on
the water and then, in fall, leave the surface and overwinter
underwater.

Highlights of the Life Cycle

Whirligig beetles are great fun to watch in their adult phase
as they move about the water with various types of locomotion. Their movement itself is mesmerizing, but with a

Whirligig beetle larva, Dineutes

closer look you will see various types of movement and other behaviors associated with them. The larval stage is generally not seen. The adults can be seen in fall and spring, but not in midsummer, when only the larval phase exists.

How to Find Whirligig Beetles

Pick a warm day in early spring and go to the quiet edge of a pond, lake, or backwater of a river and look for shiny, almond-shaped black beetles gyrating about on the surface of the water. They will be from one-quarter to one-half inch long. Sometimes they sit quietly in small groups during the day, but if disturbed will immediately start to swim about.

What You Can Observe

TYPES OF MOVEMENT

Whirligig beetles are beautiful to watch as their sleek, shiny bodies zip about through the surface of the water. It is hard to get any closer than a few feet from the beetles, for they are wary and fast, but still there are several intriguing features to their movement that can be enjoyed from a distance. There are at least three distinct types of movement. The most common movement is a slow push that takes them only a few inches. It produces no visible ripples. Another type of movement occurs when the beetles are disturbed, such as when you approach too close or when a pebble is dropped near them. Here they have an almost continuous movement for which they are named: they whirl around in small arcs and remain in a tight group without hitting one another. This movement creates a wake of ripples in front of the beetle. A third type of movement is jerky and creates concentric ripples with the beetle at their center. This is the least common of the three types.

THE BEETLE'S STRUCTURE

There are three parts of the beetle's structure that are particularly interesting. They are its feet, its eyes, and its antennae. The front pair of legs is the only one easily seen. These legs are thin and well suited for grasping prey on the water. The second and third pair of legs are much shorter but are also greatly flattened and ideally suited for swimming. The beetle swims with half of its body above the water and half below the water. It needs to be able to see predators and prey both from above and below, and its eyes have the remarkable adaptation of each being split into two parts, half adapted for sight below water and the other half for sight above water. In addition to watching for predators, the beetle is sensitive to ripples on the surface of the water. The antennae are held on the water surface, and an organ at their base, called Johnston's Organ, detects the slight changes that the ripples make in the angle of the antennae. Through this sensitivity the beetle is able to tell where other

Closeup of beetle swimming

beetles are or where other insects that it might eat are struggling on the water surface. The ripples from its own movement may also bounce back from objects in the water that they hit, and the antennae could pick these up and help the beetle avoid hitting these objects. In this way, the beetle's ripples, along with its antennae, create a possible location system using "echoes" on the surface of the water, much as a bat uses sound waves in the air.

SPRING AND FALL GROUPS OF BEETLES

As the beetles emerge from hibernation, they form small aggregations on the water surface. These groups of beetles may contain one or several species. Although you will not be able to distinguish species from a distance, you may be able to recognize the two main genera of whirligigs, for they differ in size: *Dineutus* are three-eighths to five-eighths of an inch long, and *Gyrinus* are only one-eighth to one-quarter of an inch long.

Whirligig beetles with water lily

Mating takes place in these spring groups. The male darts onto the back of the female and holds on to her while transferring sperm. The two remain together for a few minutes to many hours. After mating, the females lay eggs. Within a few weeks most of these adults have died.

New groups of adults will not be seen until late summer. These also form aggregations, which, on some larger lakes, may contain as many as 200,000 beetles. These remain inactive during the day, but at sunset the beetles disperse as they forage throughout the night, gathering insects off the water surface. Then, just before sunrise, they again form large groups, often in the same places as the day before. Sometimes individual beetles may fly up to two miles away at night to join other groups in new areas.

JUNE BEETLES

Relationships

June beetles are a genus (*Phyllophaga*) of insects in the family Scarabaeidae, which in turn is in the order Coleoptera, or beetles. Scarabs can be distinguished from other beetles by their antennae, which have little flattened projections or platelets at their tip, which can be closed and opened like the pages of a book. June beetles can be dis-

Typical June beetle habitat

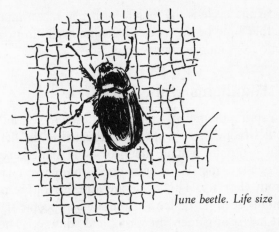

June beetle. Life size

tinguished from many other scarabs by having only three platelets at the tips of their antennae. Most other genera have more.

Life Cycle

June beetles overwinter as adults in underground burrows. In early spring they emerge, feed on leaves, mate, and lay eggs. Eggs are placed one to four at a time in small clods of earth, sometimes up to seven inches underground. The larvae hatch in two to four weeks and feed on vegetable matter in the soil. In fall they burrow down sixteen to twenty inches, where they spend the first winter. In spring they return to near the soil surface and resume feeding. In fall they again burrow down, and then in spring again burrow up and feed. In this second summer, they pupate underground in cells created by pressing the earth out around them. In a few weeks the adults emerge from the pupae but remain under the soil through the third winter and then emerge above ground the next spring. This three-year life cycle is the length for the most common species. Other June beetle species have life cycles from as short as one to as long as four years.

Highlights of the Life Cycle

In spring the freshly emerged adults come crashing against our window screens at night, attracted to the lights inside. They usually remain there for quite some time and this provides a marvelous opportunity to observe a large beetle from close up. The larvae are also one of the most commonly seen beetle larvae, found when people start digging in lawns or gardens in spring and summer. They are the most common white grub in the soil.

How to Find June Beetles

The best way to see June beetles is to wait for them to come to you. They are attracted to lights at night, and you will hear them crash into your screens or windows and then make a racket as they buzz and crawl about. They sound like fierce creatures but in fact are harmless to us. They range in length from a half inch to a whole inch and are usually brown. The larvae feed on roots a few inches underground. They are white, somewhat transparent, have light-brown hardened heads, and are usually curled around into a C. They can be found as you dig in a garden or in just about any shovelful of soil taken from a lush meadow.

What You Can Observe

ANTENNAE AND WINGS

As June beetles crawl about on our screens or windows, we can get close enough to them to see some interesting aspects of their structure, especially their antennae and wings. All scarabs can be recognized by the special form of their antennae: at the tips are several flat plates that can be opened or closed much like the pages of a book. These are small, but if you look closely, they are easily recognized. June beetles have only three leaflets at the tip of each

Extended wing of June beetle

antenna and they occur at right angles to the rest of the antenna. This feature distinguishes them from many other scarabs.

As the beetles land against the screen, you will also get a chance to observe the fascinating structure of their wings. The front pair of wings on all beetles has evolved into hardened shells that neatly close over the back of the insect. The second pair of wings on the scarabs and many other beetles, such as the ladybird beetle, are much longer than the front wings and are actually folded back up under the covering front wings. To fly, the beetle has to lift the front wing covers, expand the hind wings, and then take off. When it lands, it often takes it up to a minute to refold the hind wings under the front covers; the wings actually bend in half. You may also find dead beetles at the base of windows, especially in midsummer, when the beetles normally die, and you can then lift the front covers and pull open the hind wings to see them more clearly.

One of the reasons the beetles crash into our screens rather than landing gently is that in the process of evolving the hardened front wings that give them good protection from predators, they have sacrificed much of their flight mobility. They can do nothing other than crash-land. In many cases, the females have only very short wings and are flightless; therefore, the ones we find plowing into our windows are mostly males.

DAILY CYCLE

Most species of June beetles, in the adult phase, spend the daylight hours buried a few inches under the soil. At dusk they burrow up to the soil surface and emerge over a one to two hour period. Throughout the night they feed on the leaves of trees and other plants. Mating and egg-laying also occur at night. To attract a mate, the female gives off a scent that affects males in a twenty-yard radius almost immediately. They stop feeding, fly to her, and mate. Once fertilized, the female starts to lay eggs, either singly or in groups of three or four, placing them inside little clumps of soil up to seven inches deep. As dawn approaches, the beetles burrow back into the soil, although this time, instead of doing it gradually over several hours as when they emerged, all beetles leave feeding and enter the soil almost simultaneously, within a ten-minute period. A few species of June beetles reverse this behavior and are active in the day and under the soil at night.

THE LARVAE

The larvae of June beetles are the curled, semitransparent white grubs that you find a few inches under the soil. They

Left: *beetle grub*; right: *beetle pupa*

are found in vegetable and flower gardens but are more common in pastures of timothy and bluegrass where there are abundant sources of roots, which are their main food. There are other types of white grubs in the soil, but those of June beetles can be distinguished from them by the double row of hairs at the hind end of the body.

The beetle larvae are very destructive to agricultural crops, especially in the South.

WILLOW AND COTTONWOOD LEAF BEETLES

Relationships

These beetles are members of a subfamily (Chrysomelini) of insects in the family Chrysomelidae, or leaf beetles, which in turn is in the order Coleoptera, or beetles. There are over fourteen hundred species in this family and many are very common. Most species are less than a half inch

Leaf beetle: larvae, adult and eggs,
Chrysomela *sp. Life size*

Willow leaf beetle and cottonwood leaf beetle habitat

long, are usually oval, and have antennae no longer than half the length of their bodies. In all stages of these beetles' lives they feed on plants.

Life Cycle

The beetles overwinter as adults hidden among the leaf litter at the base of their food plants. In spring they emerge and, after a week or two of eating, mate and lay eggs in clusters (of about fifty) attached to leaves. In several days the eggs hatch and the larvae feed on the leaves, at first in groups and later singly. They mature in about a week and then pupate. The larvae pupate attached to the leaves. Pupation lasts about a week and then the adults emerge from the pupae and feed on the leaves. There may be up to five broods, the adults of the last brood crawling down into the leaf litter and overwintering.

Highlights of the Life Cycle

These are marvelous insects to observe because all of their life stages can be seen at the same place: on the leaves of the food plant. They afford a rare chance to observe the diversity of form that occurs in complete metamorphosis. The beetles are most obvious in spring because they are among the first to attack the new leaves.

How to Find Willow and Cottonwood Leaf Beetles

The easiest way to find these beetles is to first look for the plants they like to eat. They are most common on willows, either shrubs or trees, and these can be found in wet areas

Adult leaf beetles. Left: *cottonwood leaf beetle*; right: *willow leaf beetle*

such as the edges of ponds or streams or in wet ditches along roadsides. Once you have found the plants, look for the small irregular holes in the leaves created by the adults as they feed. There are three common types of these beetles to look for. Two, the willow leaf beetle (*Chrysomela interrupta*) and the cottonwood leaf beetle (*C. scripta*), are closely related and similar in appearance. They are about a quarter inch long and have nice patterns of black lines and dots on a generally yellow background. The other, the imported willow leaf beetle (*Plagiodera versicolora*) is only one-eighth inch long and is a shiny steel blue.

What You Can Observe

THE BEETLES AND THEIR SIGNS

No sooner do leaves begin to appear on trees and shrubs than insects emerge as well and start to eat them away. This seems to be particularly true of willows, whose leaves are among the first to expand in spring, for within a few weeks they are riddled with holes. In early spring the eaters of willow leaves are most likely beetles in the family Chrysomelidae. They are often called leaf beetles because so many of them eat leaves. The three beetles mentioned here feed on any member of the willow family, which includes willows, cottonwoods, poplars, and aspens. They can also be found on alders. I find them most often on alders and the shrub varieties of willows.

The first stage of eating appears as small, irregularly shaped holes in the leaves. These are made by the adult beetles. After about two weeks of feeding, the adult beetles mate. You may see pairs of them together, the male on the back of the female. After this you can look for the eggs on the leaves, for they are laid in tight clusters of about fifty eggs each, most often on the underside of the leaf. The larvae hatch in about a week, and you can tell if they have been feeding, for they "skeletonize" the undersides of the leaves, eating just the bottom surface of the leaf between the veins, leaving the skeleton of the veins. Later, as they mature, they eat holes just like the adults. The larvae are black and look a little like minute lizards, except they have three pairs of legs. The larvae of the two larger species have two small white dots on each side of their thoraxes. When the larvae are disturbed, they may lift their abdomens and/ or emit a milky, distasteful juice from around the dots on their thoraxes.

The full-grown larvae pupate while attached to the leaves, so if you see holes in the leaves but can find neither adults nor larvae, then look for the pupae. They are small black cases a little shorter than the mature larvae and are attached by their ends to the leaves.

WHITE AND SULPHUR BUTTERFLIES

Relationships

Whites and sulphurs belong to a family (Pieridae) in the order Lepidoptera, or moths and butterflies. Members of the Pierid family are easy to distinguish from other butterflies for they are our only butterflies that are either all white, yellow, or orange, often with the addition of some slight black markings. They are medium-sized butterflies with a wingspan of about an inch and a half. Their larvae are hairless and usually green with a yellow line down each side. They feed singly on mustards and legumes.

White and sulphur butterfly habitat

White butterflies (Pieris rapae) *on wintercress. Life size*

Life Cycle

Whites and sulphurs overwinter in the pupal phase, which for butterflies is called a chrysalis. In early spring the adults emerge from pupae, feed on nectar, and mate. The fertilized females lay eggs singly on the leaves of food plants, usually a mustard or legume, and the eggs hatch in about a week. The larvae mature in two to three weeks and undergo four molts during this time. They pupate attached to plant stems, and pupation lasts about ten days. There are three or more generations per year, the greater number occurring in the South. The pupae of the last generation overwinter.

Highlights of the Life Cycle

These butterflies emerge early in spring and are easily spotted as they fly above the meadows. They have some very interesting mating behavior that includes aerial displays called spiral flights. It is also fun to watch the behavior of females as they lay eggs. The larvae are inconspicuous and hard to find.

How to Find Whites and Sulphurs

To find these butterflies, simply go out on a warm, sunny day to where there are meadows or lush weeds and look for white or yellow butterflies that have about an inch and a half wingspan — these are undoubtedly whites or sulphurs. They can be easily spotted even up to a hundred yards away.

What You Can Observe

SPIRAL FLIGHTS

It is always a pleasure to stare out over a lush meadow and watch the ever-present white or light yellow butterflies as they alternately flap and glide in the warm breezes, sometimes spiraling up into the air in pairs. The whole scene seems to epitomize our feelings of warm weather's ease. For the butterflies, though, it is far from ease. The spiraling into the air is an instinctive behavior associated with mating, and even a brief study of it shows butterflies' lives to be far more regimented than we would probably ever like to believe.

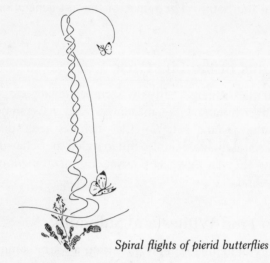

Spiral flights of pierid butterflies

Spiral flights start with two butterflies flying close together near the ground. Then one starts to circle around the other, and this is immediately followed by both butterflies spiraling around each other as they rise to sometimes as high as sixty feet. Suddenly one of the butterflies drops like a dead weight to the ground while the other slowly drifts down, over the period of several seconds.

Spiral flights are extremely common, and over a large field you may see several pairs spiraling at once. They can be seen throughout most of the spring and summer. They involve a male and a female. The female has usually already mated, and this behavior is believed to be her attempt to shake off a courting male and get back to her business of egg-laying. She is the first to rise in the spiral flight and the male follows. After reaching a certain height, the male gives up and drops straight down. The female comes down slowly, possibly to allow more time for the male to move away.

EGG-LAYING

Once mated, females spend a great deal of their time laying eggs. They lay their eggs singly on the leaves of their host plant. With a little practice you can begin to tell when a female is in the process of egg-laying and then even go and see the egg. To locate an egg-laying female, first look for a butterfly flying lazily about over plants but not stopping to feed at flowers or chase after other butterflies. This is likely to be a female looking for the food plants of her species. Nobody knows exactly how the female distinguishes between plants, but it is fun to watch. Sometimes she just comes near a plant, at other times she alights for only a moment. When she has found the right species, you will see her land on a leaf, bend her abdomen down, and briefly touch the leaf. This is all there is to her egg-laying, and if you look under the leaf you will see a small white dot the size of a pinhead.

Right: *two larvae*; left: *chrysalis*

The adult butterflies usually remain where their larval food plants are abundant. Larvae of the whites eat mostly plants from the mustard family, such as cabbage, broccoli, and wild mustards. Larvae of the sulphurs eat plants from the legume, or bean, family, such as alfalfa, clover, and wild senna. Because of their choice of food plants, both larvae and adults frequent our vegetable gardens as well.

MATING BEHAVIOR

You can also look for males that are seeking out receptive females to mate with. These males fly about fairly rapidly and approach most other butterflies. It is believed that many butterflies cannot recognize their own species until they are within about a foot of the other butterfly. Most of the time the searching male gets close to another butterfly and then moves away, either because it was the wrong species or the wrong sex. When it finds a female of its species, it starts to circle about her. If she is receptive, she quickly lands, keeping her wings closed and bending her abdomen downward. The male follows, lands next to her, attaches the tip of his abdomen to the tip of hers, and then crawls around so that the two are facing in opposite directions. They may remain paired like this for up to an hour, and may fly for short distances while still attached.

Some females are already mated and are egg-laying when the male approaches them, and they are unreceptive to his advances. An unreceptive female responds to a courting male in one of four ways: she may avoid him by crawling around to another part of the vegetation, or flap her wings vigorously to keep him from landing, or she may assume the rejection pose with her wings spread and her abdomen bent sharply upward. This last pose is, in a sense, the opposite from the mating pose, in which her wings are closed and the abdomen is bent down. If none of these behaviors stops the male's attempt to mate, then the female starts on a spiral flight. These flights are the most costly in energy to the female and therefore are her last resort to stop male courtship.

MOURNING CLOAK AND TORTOISE SHELLS

Relationships

The mourning cloak, American tortoise shell, and Compton's tortoise shell are members of a genus (*Nymphalis*) of insects in the family Nymphalidae, which in turn is in the order Lepidoptera, or moths and butterflies. This is our largest family of butterflies and its members are sometimes called four-footed butterflies. Like all insects, they have six legs, but their front pair is greatly reduced and held against the body, making them appear four-legged. Most members of this family are large, colorful butterflies.

Mourning cloak sipping sap from broken red maple twig.   Life size

Mourning cloak habitat

Life Cycle

The life cycles of these three butterflies are similiar. All overwinter as adults hidden in crevices of bark or rock. They emerge on the first warm days of spring and feed on sap from tree wounds or nectar from early flowers. After mating, the females lay clusters of eggs on their food plants, which are willow and poplar for the mourning cloak and Compton's tortoise shell, and nettle for the American tortoise shell. The eggs hatch in one to two weeks and the caterpillars (larvae) feed together. After four weeks of feeding and four molts, they break up into more scattered groups. They then leave their food plant and individually find protected spots in which to pupate. Pupation lasts about two weeks and then the insect emerges as an adult. There are two broods in the North and three in the South. In fall, the adult butterflies from the last summer brood overwinter in protected spots.

Highlights of the Life Cycle

One of the most exciting times in the lives of these butter-flies is in late winter or early spring when they emerge from hibernation and we can see them flying about long before leaves or most flowers have come out. It is also great to watch them feed on the sap from recently cut trees or broken branches. Later in spring, when the caterpillars have hatched, they are also obvious because they feed together.

How to Find Mourning Cloaks and Tortoise Shells

There is no place to go and be assured of finding one of these butterflies, for they frequent a variety of habitats and are extremely mobile. To maximize your chance, take a walk through a deciduous wood on one of the first warm days of spring (60° F. or better). The butterflies often over-winter in these areas and can be found feeding on sap from broken tree limbs or stumps. These are the earliest butter-flies to become active, so just about any large butterfly you see in late winter or early spring will be one of these three.

What You Can Observe

SPRING BEHAVIOR

If you walk through open woods on a warm day in late winter or early spring, you are likely to see, gliding among the sunlit trees, a large butterfly, its wings deep purple-brown and edged with yellow, its flight pattern a few rapid flaps alternating with strong glides. This is the mourning cloak, the first butterfly to be seen in spring. A week or two later it is joined by the Compton's and American tortoise

shells. They have wingspans of about two inches (as op-
posed to the three inches of the mourning cloak) and the
upper surface of their wings is mostly orange and brown.
One other common early spring butterfly is the spring
azure, a tiny blue butterfly unrelated to the mourning cloak
and tortoise shells.

These butterflies have similar spring habits. Amazingly,
they all hibernate through the winter as adult butterflies.
In fall the adults crawl into protected places, such as under
loose bark. In spring they emerge from these places and fly
about in search of food. Their main food is nectar from
early-blooming shrubs, such as willows, and sap from tree
wounds, such as broken twigs. They particularly like sap
with sugar in it, like that from maples or birches. A newly
cut stump from one of these trees may attract five to ten
butterflies — a truly wondrous sight. These cut trunks also
attract many other insects at this time of year, especially
flies and beetles and sometimes a few solitary bees.

The butterflies usually keep their wings open when they
are feeding, and this enables you to see the beautiful colors
on the upper surface of the wings. But if you frighten them
they often fly up and land on tree bark with their wings
closed. The coloring on the underside of their wings looks
just like the bark, and with their wings folded the insects
become almost invisible. Usually when they are feeding,
you can approach to within five or six feet without scaring
them, and from this distance you can see their long, thin
mouthpart uncurled from beneath their heads and extended
to suck up the sap or nectar.

The spring mating behavior of these butterflies is best
seen in the mourning cloak. It takes the form of short
chases between butterflies, sometimes resulting in spiral
flights as described for the white and sulphur butterflies.
Two butterflies spiral up into the air twenty to thirty feet,

American tortoise shell and
spring azures at sapling stump

and then one suddenly drops to the ground while the other glides down slowly. The mating behavior of mourning cloaks has not been well enough studied to know the exact function of these flights, but a good guess would be that they have a function similar to the spiral flights of the white and sulphur butterflies (see White and Sulphur Butterflies, spring section).

The butterflies may be very common at one time of the day and be impossible to find later. This is because as soon as it starts to get too cold for them to be flying about, they seek out places similar to where they overwintered and crawl back into them to be partially protected from the cold.

By late spring there will be very few of these adult butterflies flying about, for they will have already mated, laid eggs, and died.

BROODS

Most of these butterflies are double- or triple-brooded, the greater number of broods being in the warmer climates of the South. But there is an interesting possibility, at least with the mourning cloak, that in the North they may be only single-brooded. It is believed that the new adults that

Mourning cloak larvae and chrysalis

mature in early summer from eggs laid that spring go into a summer hibernation, called aestivation, and reemerge in fall to feed, and then go into hibernation for the winter. This would mean that they live for about ten months. The purpose of the aestivation is unknown, but it could be simply to save the butterfly from unneeded exposure to predation and wing wear.

A number of parasites attack the mourning cloak. One is a small fly, *Telenomus graptae*, that completes its whole life cycle within the eggs of the butterfly. Chalcid wasps, ichneumon wasps, and tachinid flies are also all known to parasitize the larvae.

TENT CATERPILLAR

Relationships

The tent caterpillar is a species (*Malacosoma americanum*) of insect in the family Lasiocampidae, which in turn is in the order Lepidoptera, or moths and butterflies. Larvae in this family typically live as a colony and often make web nests. The adults are light brown, stout-bodied moths with a wingspan of about one and a half inches. They lay their eggs in clusters surrounding the twigs of their food plants.

Tent caterpillar habitat

Tent caterpillars, beginning web nest, and egg case.
Life size

How to Find Tent Caterpillars

In early spring, when leaves are just emerging, walk along the edge of a wood or road and look in the forks of shrub or tree branches for the shiny, white web nests of the caterpillar. They are the only such web at this time of year. If you see none, come back a week later when the webs are larger and more conspicuous. The caterpillars will be either in, on, or near the web.

What You Can Observe

FINDING EGG CASES

You can begin your observation of tent caterpillars long before the nests are even present. Start in winter or early spring and look for the insect's eggs, which are easily found along the twigs of black cherry, chokecherry, or apple. They are a little less than an inch long and are dark, shiny masses that surround the twig. There may be from one to as many as twenty egg cases on a tree. The shiny outer surface of the egg case is water resistant. If you scrape away a little of it you will see a hardened bubbly substance inside. Beneath this is a single layer of two hundred to three hundred eggs surrounding the twig.

The caterpillars hatch from the eggs on the first few warm days of spring. They feed on the emerging leaves and buds of their tree. About a week after hatching, the caterpillars start to build a web nest, and at this point there are many features of the caterpillars' behavior to observe: building the nest, congregating, feeding, and laying down web trails, molting, avoiding parasites, and looking for places to pupate.

Life Cycle

Tent caterpillars overwinter as eggs. In early spring, the larval phase begins when the caterpillars hatch from the eggs. During the next six weeks, they feed on leaves, build their communal web nest, and molt five times. They then leave the web and individually seek out places to build a cocoon and pupate. Pupation lasts three weeks. The adults emerge in early summer, and mate and lay eggs by midsummer. The eggs remain unhatched until the following spring.

Highlights of the Life Cycle

Although all phases of the tent caterpillar's life can be seen, the easiest to find and the most entertaining to watch is the larval phase. This starts in spring when the leaves of the caterpillar's food plants, primarily apple and cherry, are just emerging. The caterpillars are active and have a number of interesting habits that can be readily observed.

NEST CONSTRUCTION

All the caterpillars hatching from an egg case help in building the nest. It is always made in the crotch of two or more branches. The nest is composed of tough layers of silk with spaces in between each layer. There are always one or more openings to the nest, which the caterpillars

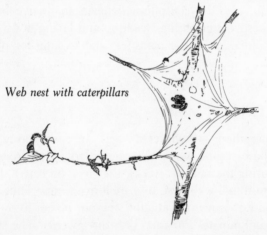

Web nest with caterpillars

use for getting in and out. On clear, warm days the caterpillars usually have two periods of nest building: one in mid-morning and the other in the early evening. Once the nest is begun, the caterpillars continue to add to it until they have molted for the fifth time and are about to leave and pupate. At this stage the nest is no longer maintained and begins to fall apart.

The nest is constructed of sheets of webbing that are

continuous around the whole nest. When a new layer of webbing is added, space is left between it and the last layer, which enables the caterpillars to crawl around within the nest. The caterpillars deposit their droppings and shed skins in the inner layers, so new layers are needed both to contain this waste material and to house the larger caterpillars. The function of the web nest is unclear. Some think it is protection from predators; others suggest that it may act like a greenhouse, creating a moist and warm environment that fosters the growth of the insects.

CONGREGATING

For reasons that are not understood, the caterpillars often congregate in a tight cluster on the outside of the web nest. They may do this to warm themselves in the sun after a cold night, or they may do it to get out of the heat and humidity of the nest on a hot day. Once out of the nest, the grouping may provide some form of protection from certain predators.

FEEDING AND MAKING TRAILS

Generally the caterpillars leave the nest to feed on leaves three times a day: once in early morning, once at midday, and once just after dark. The longest feeding period is at night. Each time the caterpillars leave the nest to feed, they leave a trail of silk on the twigs to mark the way back to the nest. If you look on the twigs you can see these quite clearly. They are thicker on twigs that have been most used. Research on these trails has shown that they relay information to other caterpillars as to where the best sources of leaves are. The silk trails of caterpillars returning from good feeding spots are the most attractive to outgoing caterpillars.

The daily pattern of the caterpillars is interrupted by two

things: cold or rainy weather and molting. Each caterpillar sheds its skin five times in its life, and surrounding each molt there is a period of inactivity of about two days. During this time the caterpillars stay inside the nest.

LEAVING THE NEST AND PUPATING

Once the caterpillars have molted for the fifth time, they crawl to the tips of the branches and either descend to the ground on a silk strand or, more commonly, flick their bodies right off the twig. At this time you are likely to see the caterpillars crawling about in search of places to pupate. Often they can be seen crossing roads or highways that are

Adult and cocoon

lined with cherry trees, and many of them get killed by cars. Once they find a protected spot, such as under a log or in a bark crevice, they weave a tough cocoon. This is only half as long as the caterpillar, so the caterpillar doubles over to fit into it. Before pupating, it excretes a yellow substance that stiffens the cocoon and dries to a powder. When you dislodge a cocoon or flick it slightly, the powder goes into the air like chalk dust. This is unique to this genus and can be used as an aid to identification. The adult moths emerge in midsummer, mate, and deposit new egg cases by late summer. They are all tan and stout-bodied, and have a wingspan of about an inch. A faint white line runs across their fore wings. After mating and laying eggs, they die.

PARASITES

If, while you are watching the caterpillars, you see a fly buzzing about near the nest, keep your eye on it, for it may be a tent caterpillar parasite. One such parasite is a tachinid fly that looks a little like a large housefly. It darts onto the nest and lays its egg just behind the head of the caterpillar, in a spot where the caterpillar cannot chew it off. The egg hatches and the fly larva burrows into the caterpillar and feeds on it. The caterpillars have evolved at least two defenses against this. They may quickly flick their heads back and forth when the fly is around to make it harder for the fly to oviposit (deposit its egg); or they may even be able to recognize the pitch of the buzz of their predator and move off the surface of the nest when they hear it.

ICHNEUMON WASPS

Relationships

Ichneumon wasps are a family (Ichneumonidae) of insects in the order Hymenoptera, or bees, wasps, and ants. They are one of the largest families of all insects. Almost all species of ichneumon wasps are parasitic, developing on or within the bodies of other insects. Some even parasitize

Where ichneumon wasps can be found

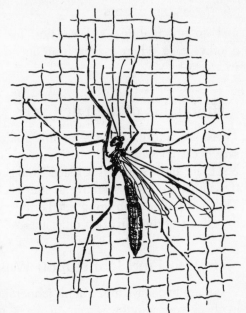

Adult ichneumon,
Ophion *sp. Life size*

other ichneumons that have already parasitized other insects. These are then called hyperparasites.

Life Cycle

The life cycle of most ichneumons is still unknown. What follows is a general account containing the main elements from the known species. Ichneumons overwinter in the pupal stage, encased in a cocoon (a few overwinter as adult fertilized females). In spring the adults emerge, mate, and look for other insects to lay their eggs in. Most ichneumons parasitize only one species; many choose a species of moth or butterfly caterpillar. The eggs hatch and the larvae develop inside the host, usually without killing it, until the wasp larvae have completed their development. Sometimes the wasps start their development in one stage of their host and do not mature until the host has transformed into another stage. There may be from one to many generations in a year.

Highlights of the Life Cycle

These are unusual insects for this guide, for it will be a rare occasion when you observe much of their lives. All but their adult stage takes place inside other insects, and the adults are often secretive as they seek out their hosts. They are included in this guide because you will often see the effects of their lives on other insects.

How to Find Ichneumon Wasps

There are two ways to see adult ichneumon wasps. One is to check your window screens at night, for several species are attracted to lights in the dark. The other way is to look for them as they hunt for hosts. I most often find them hunting around the variety of plants at the borders of woods. The next problem is to recognize an ichneumon, and I am sorry to say there is no sure way for the beginner, except to look for a thin, delicate wasp with antennae at least half as long as its body. Sometimes the antennae have white or yellow segments at the middle. They range in size from minute to one and a half inches long. The species attracted to screens is orange-brown and about an inch long.

What You Can Observe

WASPS AT HOUSE SCREENS

Although there are over six thousand species of ichneumons in North America, and although the wasps are extremely common and abundant, there are very few that come into contact with humans. One group that does come into view is attracted to lights and often enters houses in late spring. Its species are orange-brown, about an inch long, and have long antennae. (I have handled these ichneumons with varying results as I try to get them out of the house. Usually they do not bother me as I lightly grab them in my hand, but once one poked my hand with its sharp ovipositor and it stung considerably.) Their abdomen is long, slightly curved, and flattened on each side as if it were pinched. This is the genus *Ophion*, which lays its eggs on caterpillars such as the cutworm, one of several types of moth larvae that infest our garden and mow down the first green shoots. The wasp larvae mature inside the cutworm, finally killing it. There are several species of this same group that parasitize the larvae of silk moths, tussock moths, tent caterpillars, and fall webworms.

THE LARGEST ICHNEUMON

One of the largest ichneumons is *Megarhyssa macrurus*. Its body is two to three inches long and its long ovipositor adds two to four more inches. The ovipositor is obvious when the insect is in flight and looks like several strands of thread trailing behind. It is used to penetrate through wood and lay an egg in the developing larva of a horntail, a primitive wasp whose larva feeds in tunnels inside the wood. It is believed that the ichneumon can sense the vibrations of the feeding horntail larvae with its antennae. The egg develops in the larva, not killing it until it is full grown. The wasp pupates in the horntail tunnel, and, when an adult, chews its way out through the bark. In splitting wood you may find the mature wasp or pupa inside the tunnels of the horntail. The adult *Megarhyssa* is most common in late spring.

You may even be lucky enough to watch one of the

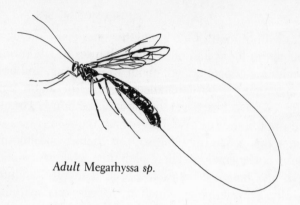

Adult Megarhyssa *sp.*

females deposit eggs in a tree. She first moves about tree bark, rapidly vibrating her antennae against it, then stops at one spot and starts to bend her ovipositor up and back toward herself. When it is looped over, she stands higher and higher on her legs until the ovipositor is going straight into the bark. Then, in some amazing way, she is able to insert it into the tree, possibly taking advantage of cracks. She may lay eggs several times in one area of the tree. Males are often in the area or on the same tree, possibly looking for females to mate with.

HUNTING BEHAVIOR

As you look over flowers and leaves in search of insects, you are likely to encounter wasplike insects with long antennae that are neither feeding at flowers nor eating leaves but moving about and feeling with their antennae. It is very likely that these are ichneumon wasps searching for their prey. Many of our larger ichneumon wasps lay their eggs on caterpillars, and caterpillars are most often found where there are lots of leaves to feed on. To me, it is always wondrous to watch one of these ichneumons at work, for it knows some way to find and distinguish its host, which is probably some species I never even knew existed.

HOW TO TELL AN INSECT HAS BEEN PARASITIZED

When the host insect is still living and active, there is no way to tell if it is parasitized by just looking at it. However, you may come across dead caterpillars or cocoons that look dry, brittle, and no longer viable; these are often the signs of the work of ichneumons. To tell if an ichneumon was there, look through the remains of the caterpillar or cocoon and see if you find smaller, hard, oval pupal cases inside. These are signs that the ichneumons matured inside the host and, after finally killing it, made their own pupal cases. The cocoons of the large silk moths are the places where ichneumon pupal cases are most frequently found by the average observer.

ANTS

Relationships

Ants are a family (Formicidae) of insects in the order Hymenoptera, or bees, wasps, and ants. We think of ants as being wingless, but during a short part of their life cycle, when they mate and disperse, they have wings. It is because of the structure of these wings that ants, bees, and wasps are related.

All ant species have societies with several castes. There are many species and they are difficult to distinguish.

Ant habitat

Ants, Formica *sp. Life size*

Life Cycle

Ants overwinter as adults in a colony that consists of a queen and her workers (sterile females). In spring the queen resumes laying eggs and the workers tend the eggs and gather food for the whole colony. The eggs hatch into larvae, which in turn pupate. The length of these three stages varies with the species, but in some of our common species each stage lasts about three weeks. Sometime in spring or summer the queen lays eggs that develop into winged males and winged fertile females. These winged ants leave the colony, fly about, and mate. After this, the males die, but the fertilized females, or new queens, seek out little cavities under stones or bark, seal themselves off from the outside, and lay eggs. Without ever leaving her new nest, the queen raises these ants to maturity, feeding them with her saliva and other eggs that she lays. It may take her up to nine months to raise this first brood, but once they are mature, these workers leave the colony to retrieve food for the queen and future broods. Colonies gradually increase in size over the years. Individual workers may live as long as six years and queens as long as fifteen years.

Highlights of the Life Cycle

Adult ants can be found in any season, but they are most conspicuous in spring when they are enlarging or excavating nests, engaging in territorial battles, or doing mating and dispersal flights. These group behaviors are particularly exciting to watch, but equally interesting are things individual ants do, such as foraging, following trails, and cleaning themselves.

How to Find Ants

There are several good ways to find ants besides having a picnic. One is to look for small mounds of excavated earth, in dry dusty areas; these are at the entrances to nests. Another way is to look under stones or logs, always being careful to replace them after you have looked. The larger group activities of ants will be found mostly by chance.

What You Can Observe

TYPES OF NESTS

The main function of ant nests is to provide a protected and controlled environment in which to raise the young. Ants are extremely adaptable in their nesting habitats and will take advantage of a variety of natural or man-made cavities to meet their needs. Many queen ants start nests under stones, for these absorb heat from the sun and slowly give it off during the night, keeping the nest an even, warm temperature. Once the first brood of workers matures, they excavate the earth beneath the stone and either carry it away from the nest or drop it just at the nest entrance, thus creating a small mound or crater. These craters are conspicuous along sidewalks and in open dirt areas.

Another type of ant nest is created when the workers gather material from outside the nest and form a mound above the earth with tunnels throughout it. These can be from one to several feet in diameter and are usually slightly soft when stepped on. The rounded shape of these mounds catches more of the sunlight than a flat surface, making the interior of the mound up to 20° F. warmer than the surrounding air.

Other nests are in plant cavities that are either already there or created by the ants. Some ants are known to live in hollow oak apple galls (these look like tan Ping-Pong balls attached to twigs); others live in empty seed cases; and still others, such as the carpenter ant, carve away the wood to make their own tunnels.

It is hard to identify ants by their nests since related species often have very different nesting habits, and even the same species may change their nest structure as the colony gets larger.

ANT SOCIETY

When you pick up a stone or log and see ants swarming in all directions, chances are you have uncovered a nest. As you have probably noticed, there are never simply adult ants in ant nests. The adults may be of several different sizes; some may even have wings. Then there are also a number of little white things that the ants often carry

Ant nest with eggs, larvae, and adults

around. What you are seeing are the various types of adults in the species and also all of the developmental stages of the ants.

You may have read or seen films on types of ant societies in which there are many castes, each doing a very specialized job. These species tend to get all of the publicity, but they make up only about ten percent of the common species around us. The vast majority of ants have a relatively simple society, with a queen, which is usually about twice the size of the largest worker, workers, and then males and fertile females. The queen and the workers are wingless and are the types of ants most often found in the nest. The winged males and winged fertile females are only in the nest for a week or two before they leave and disperse.

The little white things in the nest are the developmental stages of ants. The eggs are round and usually stuck together in bunches. Sometimes they are too small to be seen. The larvae are whitish and oblong and vary in size from just larger than the eggs to just smaller than the pupae. They usually are shiny and glistening. The pupae are a little smaller than the adults. They are in white cocoons whose surface is dull and parchmentlike.

LARGE GROUPS OF ANTS

There are two occurrences you will see that involve large groups of ants. The most common is swarms of ants flying about, landing on buildings, or crawling on the grass. These are not termites, for adult termites are white, but rather the winged males and winged females emerging from the nest. They will fly about and mate, and after mating, the new queens either shed or bite off their wings and so are wingless from then on. Often you can see the wings lying about. Each queen attempts to start a colony by first finding a suitable nest site, then laying eggs, and raising the young until they mature to adults.

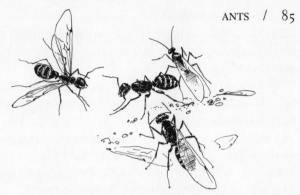

Dispersing carpenter ants: three fertile females in various stages of shedding wings, and male in upper right

The other situation is where hundreds of ants seem to be crawling all over each other. This is often seen on sidewalks. In this mass, there may be ants of two sizes or ants all the same size. In the first case, there are most likely two species, and in the second, most likely one. These masses are probably involved in an aggressive encounter, possibly over the position of nest sites. They could be termed "territorial battles" or even "wars." In contrast to our wars, they are conducted entirely by females. If you look closely at the ants, you will see individual battles — ants using their pincers to dismember the bodies of other ants. On the battlefield may be cutoff legs or heads. Sometimes several ants will gang up on another and often the encounters are like quiet tugs of war rather than fierce scrambles.

WATCHING TWO ANTS MEET

About five levels of interaction have been observed between different worker ants and you may be able to observe many of these when you see two ants meet. Two interactions commonly occur between ants of the same colony. The first is called examining: one ant directs the tips of its antennae over the surface of another ant. A closer interaction is called licking: one ant touches its mouth onto the body of the other ant. The ant that is being licked often becomes still while this is occurring.

The next three movements are associated more with aggression and most often occur between members of different colonies of the same or different species. The mildest aggression is threat. In this, an ant raises its head and opens its mandibles or pincers. The next level of aggressive action is seizing, where one ant seizes a limb or body section of another. A further aggressive action is termed dragging, where one ant drags another after seizing it.

TRAIL FOLLOWING

Once an ant leaves the nest, it has the problem of finding its way back. Ants are believed to do this in two ways, with visual clues and with chemical clues. Ants that use visual clues, such as a prominent part of the landscape like a tree or rock, or the polarized light of the sun, tend to travel in straight lines to and from the nest. Other ants of the same colony will travel parallel to it but not in a line behind it. Instead of visual clues, though, most common ants seem to use scents called pheromones to mark a trail on the ground that they can then follow back to the nest. This is what is happening when you see several ants all going along the same path. If you rub your finger across this path, you will see that the ants stop where you rubbed, for you have removed the pheromone from that spot. Not until several ants explore the spot will the pheromone be sufficiently reestablished for the ants to continue freely along the path. Pheromone-marked paths can be more winding, and many ants will travel the same path in follow-the-leader fashion.

FORAGING

Once an ant finds food, it generally takes it back to the nest, but if it is too big, the ant returns to the nest without the food and more workers go out to try to bring the food item back. Seeing a number of ants cooperate in bringing a caterpillar, dead cricket, or bread crust back to the nest is fascinating. One observer has noticed that ants get better at pulling their prey over a period of about ten minutes, at which point they are pulling it at their maximum rate. This learning to pull is roughly divided into three stages: initial pulling by a few ants, a slowed phase when more ants are pulling but in opposing directions, and finally an efficient time when all ants are pulling in concert. A few common ants do not drag large prey but carry back to the nest little bits that they have torn off.

ANTS CLEANING THEMSELVES

One of the main things that lone ants do is clean themselves. This is fun to watch. An important part of the ant used in cleaning is a little comb along a middle joint of each front leg called the strigil. Almost all cleaning actions end up with dirt being transferred to the strigils and then cleaned off with the mouth. The antennae are rubbed through the strigil often, and while one antenna is being pulled through the strigil on one side, the strigil on the other side is being cleaned with the mouth, and vice versa. For cleaning the legs, the third pair is rubbed clean by the second pair, and the second pair is rubbed through the strigils of the first pair, and then again the mouth cleans the strigils. Watch single ants and see if you can see these actions.

PAPER WASPS

Relationships

Paper wasps are a family (Vespidae) of insects in the order Hymenoptera, or bees, wasps, and ants. There are several species of paper wasps; they are all social and all make paperlike nests in which they raise their young. Vespids can be distinguished from all other wasps by how they hold their wings when at rest. They are folded and held out to each side rather than being folded over the back as in other

Queen wasp starting new nest. Life size

wasps. Even though the family Vespidae contains our only social wasps, most of its members are solitary. Hornets and yellow jackets, both social wasps, are in this family as well.

Paper wasp habitat

Life Cycle

Paper wasps overwinter as fertilized queens. They emerge in mid-spring, build paper cells from collected material, and lay eggs in the cells. The eggs hatch into larvae in about two weeks. The queen feeds them for another two weeks, at which time they pupate in their cells. Pupation lasts about three weeks. The mature wasps that emerge are sterile females, and they do all caring for the young that develop from eggs the queen continues to lay. In late summer, certain young develop into fertile females and males. The social order of the hive breaks up, the sterile females stop working, the queen stops laying eggs, and the male and fertile female wasps mate. The fertilized females (now queens) are the only ones to survive the winter.

Highlights of the Life Cycle

You can watch all aspects of the lives of these wasps, but it is particularly easy to watch the queens in early spring, for they are conspicuous as they build their nests and care for their young. These wasps can sting if you approach too close to their nests, but they seem to be less aggressive in the early part of the season when the queen is the only adult in the nest.

How to Find Paper Wasps

Look for the queens building nests under the eaves of buildings, collecting wood pulp from old exposed pieces of wood, or collecting water from a muddy spot at a puddle or edge of a pond or stream. They are thin wasps, black or dark brown, with yellow or orange markings on their abdomens. Like all of the family they belong to, they fold their wings when at rest and hold them on each side of the body rather than over the back like all wasps in other families.

What You Can Observe

GATHERING NESTING MATERIAL

Queen wasps flying about in early spring are often involved in gathering materials for nest construction, and it is fun to watch them during these activities. The wasps use fibers from dry wood or paper, and these are usually gathered from dead limbs, exposed boards, discarded paper objects, or from the stems of weeds. The gathering behavior is easy to recognize, since the wasp is spending a great deal of time on one of these building materials. It grasps the fibers in its mandibles and peels off a thin strip while moving backward. This strip is formed into a ball and carried back to the nest in the queen's mouth. The wasp may have to try several locations on the material to get fibers of the right consistency.

Another type of gathering behavior can be seen near any source of water, be it a pond, puddle, birdbath, or watering

Queen at pond edge gathering mud

can. The wasp stands still at the edge of the water while it drinks. The water is taken back to the nest, possibly combined with other substances inside the wasp, and then regurgitated during nest-building. It is used to help connect the fibers of the nest, and also as a shiny, semiwaterproof coating on the top of the nest. If you see a spot where a wasp has gathered or is gathering wood or water, chances are good that it will return to the same spot soon for another load, since once these sources are found, they tend to be used repeatedly.

FORAGING

Wasps are carnivorous for the most part and feed on other insects, especially soft-bodied larval forms, such as the caterpillars of moths or butterflies. Wasps that are foraging can be recognized by their lazy, drifting flight over dried leaves or around shrubs. Every so often in their wanderings they land and crawl about on leaves or twigs. Once the nest is made and the first eggs laid, the queens mostly forage for food to feed the larvae. Successful foraging spots tend to be returned to as well. Usually the wasps are alone as they forage, whereas good sources of water or wood may attract several wasps at a time.

RAISING THE BROOD

In early spring there are only queens around, because the males, alive in fall, died during the winter. When it is still quite cool out, these queens tend to gather into clusters and remain rather motionless. When it is warmer, you will see an individual queen building a small single layer of a few cells in which she lays eggs, one in each cell. Once the eggs hatch, she feeds the larvae until they are full grown. Then she covers over the cell with more paper, and the larvae pupate and then emerge from the cells as adult females. The queen then continues to build, or at least start, new cells and lay eggs in them, while the workers finish building the cells and gather food and feed the larvae. Although the queen will still take trips from the nest, these are only for the purpose of gathering either building materials or water. Her food is gathered by the workers.

AGGRESSIVE SIGNALS

With the paper wasps, you don't have to guess whether you're in danger of getting stung, for there are some definite behaviors of the wasps that will tell you when to back off. One sign is the speed of the flight of a wasp near you. When a wasp is disturbed enough to want to sting you, it will fly in a straight line directly at you, at a very fast speed; so fast that it is hard to get away. This is rare and normally occurs only when you are too near the nest. A slow, meandering flight is usually safe, and if the wasp is coming toward you, it is usually by mistake, and a gentle wave or slow movement away will encourage the wasp to leave. If you see a wasp standing still on any object or even on the nest and want to get a closer look, there is another signal it will give you if it thinks you are too close. It raises high on its legs, lifts its wings, and raises its front legs. This is a good sign to stop coming closer.

SOLITARY BEES

Relationships

Bees are a superfamily (Apoidea) of insects in the order Hymenoptera, or bees, wasps, and ants. Over 99 percent of our bees are solitary; that is, they live alone rather than in societies, like the bumblebee and honeybee. This section deals with two of our largest families of bees, Andrenidae and Halictidae. All species from these families are solitary and nest in ground tunnels.

Life Cycle

Andrenidae. Adult males and females overwinter in underground burrows. They emerge in early spring and feed

Sandy bank where solitary bee tunnels are found

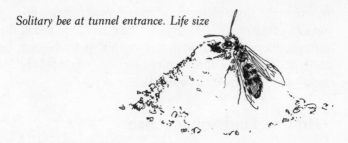

Solitary bee at tunnel entrance. Life size

on pollen and nectar from the earliest flowers. After mating, each female digs a burrow in the soil and collects pollen, which she stores in the base of the burrow. Soon, she lays an egg on the pollen mass and seals over that portion of the burrow. She repeats this process in the burrow until it is all filled with these cells, each containing pollen and an egg. She then seals off the burrow and leaves. The eggs hatch, and the larvae feed on the stored pollen. They pupate and emerge as adults by the end of summer. Some species remain in their underground cells until the following spring; others emerge in late summer and then dig a new burrow in which they overwinter.

Halictidae. Fertilized females overwinter and, like Andrenids, raise brood in underground tunnels. However, the bees that mature are all female. They, in turn, make tunnels off their mother's tunnel and raise brood without mating. These develop into males, which mate with a second brood produced by the original overwintering females. The resulting fertilized females dig burrows and overwinter in them.

Highlights of the Life Cycle

The adult phase is the only one visible, and it is most obvious in spring when the first flowers emerge. It is fun to watch the bees collect nectar and pollen from flowers and carry them to their burrows. Once you have found the burrows, there is much behavior around them that can be enjoyed. The bees rarely sting and can be approached closely and even held without danger.

How to Find Solitary Bees

You can find solitary bees in two places: where they feed and where they make their burrows. To find them feeding, look on the earliest blossoms, such as willow, shadbush, or cherry. The bees are generally smaller than honeybees, dark black with a few white or light yellow hairs over their bodies. You can also find the bees by looking for their burrows, which are made in bare earth or sand, especially where it is packed, such as along roads or paths. They are most often in the sun. The burrows are the diameter of a pencil and have excavated earth piled up around them to the height of about an inch.

What You Can Observe

BEHAVIOR AT FLOWERS

The social honeybee gets so much publicity that most people believe it is the only kind of small bee around. The fact is, the honeybee is not even native to North America, but was brought over here from Europe. Before it arrived, the only bees around were solitary bees and bumblebees. Solitary bees are among the most personable insects to watch, for they are easily and safely approached and have fascinating behavior. Amazingly, their habits are still largely unknown even by the scientific community. They look similar to honeybees but are smaller and darker. One way to be assured of seeing them is to find some pussy

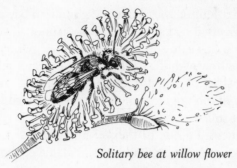

Solitary bee at willow flower

willows in early spring and look among the blooming catkins on a warm day. These flowers are one of the preferred
haunts of solitary bees in the family Andrenidae.

You may see two types of behavior among feeding bees:
those that are just feeding on the nectar and pollen, and
those that are actually collecting the pollen on their hind
legs and flying off with it. The former are likely to be
males, for they just feed themselves and, in fact, have no
place to carry pollen on their back legs. The latter are
females, for they collect pollen to store in the nest for the
developing young to feed on.

The habits of the male solitary bees are just beginning
to be studied in detail, and in many species it has been
found that the males patrol areas where the females come
to feed in the early part of the morning, and then mate
with them when they arrive. You might also look for male
solitary bees hovering near flowers and darting out after
other bees or flies that come into the area. It is believed
that these male bees are territorial, and defend the flower
site against other males and other insects that might feed
on the flowers.

BEHAVIOR AT THE NEST

The best way to find the nests is to go out on a sunny day
to a dry, open area and look for ten to a hundred bees

buzzing a foot or two above the ground. They are usually quite dispersed and so not that conspicuous. If you see any bees drop down to the earth, go to that spot and look for a little burrow with the earth mounded up around it. Often, on the sides of old dirt roads, the burrows are obvious because of the brighter colored earth excavated from lower levels of soil.

The nests are dug in the soil by the female bees. They are a foot or more deep and have a diameter about the size

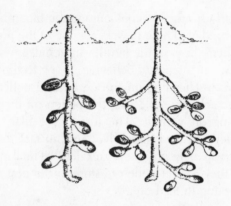

Solitary bee tunnels. Left: Andrena; right: Halictidae

of a pencil. Little clumps of earth are usually mounded up to one side. The holes are dug in dusty or sandy soils in open places, such as old meadows or the edges of dirt roads or paths. The female digs the main tunnel straight down and then creates chambers off the side of it. These chambers are about a half inch long and a quarter inch in diameter. The bee lines the walls with a substance that makes them slightly hardened and glazed. Then she brings into the chamber bits of pollen and nectar and forms them into a small, hard ball the size of a pea and lays an egg on this. She then closes off the entrance to the chamber and works on the next one.

In halictids, the first brood of females makes its brood tunnels off its mother's tunnel. This means that there may be more than one adult going in and out the main burrow entrance. In some species, one female always waits at the entrance of the burrow, possibly as a guard against predators or parasites. As you start to look at solitary bee burrows, you are likely to see this behavior, for it is common. You will see a bee with its head just in the burrow entrance, and it will move out of the way as other females with pollen enter.

OTHER INSECTS TO LOOK FOR NEAR SOLITARY BEE NESTS

Around the entrances to nests of solitary bees, there are a number of interesting parasites and predators that you should be on the lookout for. Two of them are wasps and both are very colorful. One is bright iridescent, or metallic, green. It is called a cuckoo wasp and is about the size and shape of the solitary bee, but it has no hairs on its body. If you have a group of burrows in a sandy or dirt area, scan the area and look for this wasp flying about. You may soon see it land near one of the holes, enter, and remain inside for a minute or more, and then fly out. It has just laid its egg in the chambers that the bee is busy provisioning. The wasp larva then feeds on either the stored food or the bee larva eating the food, or both. It completes its development in the chamber and emerges as an adult.

The other parasite is a wingless wasp that looks like an ant, covered with short hairs. This velvet ant is usually black or red in color. It crawls into the burrows of the bees and, like the cuckoo wasp, lays its eggs in the bee's chambers. These wasps can sting, so should not be touched.

A third insect to watch for in the area of the burrows is the blister beetle, so called because when handled it can excrete a substance that causes blisters or skin irritation. A common Northern species has very short wings, with a large abdomen sticking out from beneath the wings. It is dark metallic blue, and about an inch long. It lays its eggs in spring near places where the bees alight, such as on grass stems or flowers. The larvae attach themselves to the bodies of the bees and get a ride to the bees' burrows. When in the burrow, they detach themselves, enter the larval chambers, and mature by feeding on the food inside.

BUMBLEBEES

Relationships

Bumblebees, along with honeybees, are a family (Apidae) of insects in the order Hymenoptera, or bees, wasps, and ants. Both differ from all our other bees in being social rather than solitary. Bumblebees are large, robust bees with lots of hairs on the backs of their abdomens and with color patterns of yellow and black. They look similar to the closely related carpenter bees, except that the abdomen is densely covered with hairs, while the abdomen of the carpenter bee is practically hairless and shiny.

Life Cycle

Bumblebees overwinter as fertilized adult females — queens. These emerge in early spring and start a nest by choosing an existing underground cavity, collecting pollen into clumps, and laying eggs on the pollen. The pollen

Queen bumblebee. Life size

Bumblebee nest habitat

and eggs are covered with wax and the queen sits on them, keeping them warm while they develop. The eggs hatch in four to five days. The larvae feed on the pollen, and in about a week pupate in tough cocoons that they make. During pupation, which lasts about ten days, the queen takes off the wax that was covering them. The emerging adults are sterile females, and they take care of the subsequent broods during the summer. In late summer, the queen lays eggs that develop into fertile females and males. These leave the hive, mate, and the fertilized females overwinter. All other members of the colony die.

Highlights of the Life Cycle

One of the most interesting times to observe bumblebees is when the queens first emerge in spring, for it is then that they search for nest sites and do a great deal of foraging on the early spring flowers. Watching them closely often leads you to the nest. Other stages of their lives take place hidden underground.

How to Find Bumblebees

The queens are the largest bumblebees you will see all year, and they are very conspicuous in early spring, buzzing loudly as they fly about. You will find them in two places: at early spring flowers as they collect food, and flying low over the ground in meandering patterns in woods or open fields as they look for nest sites. It is not an insect that you can always find when you want, but rather one that is frequently encountered as you walk in the woods looking for other things.

What You Can Observe

SEARCHING FOR NEST SITES

The first bumblebees of spring always seem particularly loud and buzzy. This may in fact be the case, since these early arrivals are the queens that have overwintered from the last year's colonies. They are the largest bumblebees you will see until new queens are produced in the fall. A prominent feature of their behavior is their constant weaving flight low over the ground, interspersed with periods of landing and crawling in and out among the fallen leaves and grasses. They are out looking for places to make nests, and the most suitable spots are underground chambers, such as the old burrows of chipmunks, mice, or moles. Once they have chosen a spot, they collect grasses, moss, and leaves and form a soft ball of these materials inside the nest. Before starting their brood in the nest, they often gather more material and pile it around the outer entrance to the nest. This is believed to function in part as camouflage of the nest entrance, since suitable nest sites are in short supply and there is competition between and within species for the best sites.

This pressure for nest sites makes it advantageous to emerge from hibernation early to claim the best places. But a bee can't emerge before the flowers have bloomed to provide food, and in Northern areas snow may still cover the best sites if a bee emerges too early. It is estimated that about 10 percent of all nests are taken over by other queens. This is accomplished by a fight that either kills the original queen or makes her submissive.

PARASITIC BUMBLEBEES

There are two genera of bumblebees. One is *Bombus*, and the habits of these have been described above. The other is *Psithyrus*, and strangely, the queens of this genus have no place on their hind legs for carrying pollen. Furthermore, these queens produce only males and fertile females, but no sterile female workers. The reason for this odd occurrence is that *Psithyrus* is parasitic on colonies of *Bombus*.

Psithyrus emerge from hibernation later than *Bombus*

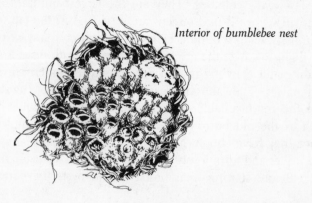

Interior of bumblebee nest

and only after the colonies of the latter have gotten well under way. They enter the nests of *Bombus* and, through various means, dominate the workers and original queens. They then lay their eggs in among the center of the existing brood, and the workers of the *Bombus* queen raise them as if they were their own. If the original queen is still in the nest and not killed, the invading queen kills any of the eggs she lays from then on. In the East, the common species of this genus is *P. variabilis*, and it most often parasitizes the nests of *Bombus americanus*.

Summer Insects

Observing Insects in Summer

Insects are everywhere in summer and this makes it both easier and harder for the insect watcher: easier to find insects, but harder to distinguish among them. Probably the best place to start looking for summer insects is in any patch of tall grasses and weeds. Two of the most abundant insects in these habitats are leafhoppers and aphids. Both are tiny insects so you must look closely to appreciate them. The leafhoppers hop around among the plants, while the aphids will be found clustered at the growing tips of the weeds. One great feature of aphids is that there are several other common insects that either attend them, such as ants, or feed on them, such as ladybird beetles, lacewings, and syrphid flies. If there are any flowers nearby, check them for adult syrphid flies; they look almost exactly like bees but can be easily distinguished with a few helpful hints (see syrphid description). The milkweed bug and tortoise beetle inhabit meadow areas as well, but they are usually on specific plants — the tortoise beetle on bindweed and the milkweed bug on milkweed.

For those interested in observing courtship and territorial behavior, the most rewarding habitats to visit in summer are the edges of ponds, lakes, or shady streams. Here you will find damselflies and dragonflies actively defending territories and vying for mates. In these activities they use several body and wing postures to communicate within their own species, and these can be easily recognized. Also in these areas you are likely to see swarms of midges. You may have always avoided these in the past, but if you read the description of them, I am sure that you will find them an endless source of fascination in the future.

Another fruitful area for insect observation in summer is the edge of a wood or forest. Here both the fireflies and robber flies are the most interesting insects to watch. Robber flies sit on perches, such as bare twig tips or certain leaves, and fly out after passing insects, trying to catch them in the air and bring them back to eat. Fireflies are active at night and have a marvelous system of flash communication that is not only beautiful but intriguing, once you begin to understand it. Also at the edge of the woods, you may see the web nests of cherry-tree leafrollers or the marvelous shapes of adult treehoppers on certain trees or vines. And finally, on the hot summer days, listen for the long-drawn-out calls of cicadas coming from the treetops. The sounds are those of the male trying to attract a female. Under the trees you may find their shed nymphal skins and the holes in the earth where they emerged from their several years of underground life.

SUMMER INSECT LOCATION GUIDE

IF YOU ARE NEAR / LOOK FOR	FIELDS	PONDS & STREAMS	WOODS	FIELD EDGES	HOUSES	BARE GROUND
WHITE-TAILED DRAGONFLY	■	■				
BLACK-WINGED DAMSELFLY		■	■			
LARGE MILKWEED BUG	■					
CICADAS			■			
LEAFHOPPERS	■					
TREEHOPPERS	■			■		
APHIDS	■					
TIGER BEETLES						■
FIREFLIES	■		■	■		
TORTOISE BEETLES	■			■		
CHOKECHERRY TENTMAKER				■		
MIDGES		■	■		■	■
ROBBER FLIES				■		
SYRPHID FLIES	■			■		

WHITE-TAILED DRAGONFLY

Relationships

The white-tailed dragonfly is a species (*Plathemis lydia*) of insect in the family Libellulidae, or common skimmers. Common skimmers are a type of dragonfly in the order Odonata, or dragonflies and damselflies. Adults in this order all have two pairs of long, membranous wings and a long, thin abdomen. For the most part, adult dragonflies and damselflies can be distinguished by the way they hold their wings when at rest: dragonflies hold them out to the sides; damselflies hold them up over their backs.

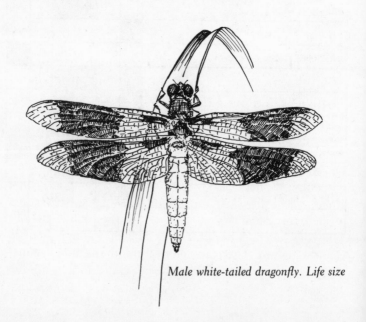

Male white-tailed dragonfly. Life size

Dragonfly habitat

Life Cycle

White-tailed dragonflies overwinter as nymphs underwater. At any time from late spring through summer, nymphs that have matured crawl two to twenty feet from the water. Then, after climbing up a plant stem, the exoskeleton on the back of the thorax splits open and the adult crawls out. Adults rest several hours, then fly to woods or fields to feed for two to three weeks. After this, they return to water, mate, and the females lay eggs by dipping the tips of their abdomens in the water and washing off the eggs as they are pushed out. After a week at the pond, most adults die off. The eggs hatch in about a week and the nymphs start to eat and mature. There are ten or more molts, and it sometimes takes nymphs more than one winter to mature, depending on the warmth of the water and the availability of food.

Highlights of the Life Cycle

All aspects of dragonfly life are fascinating to observe, but certainly the most active and engaging is when the adults return to water and mate, for it is then that males compete with each other for dominance over certain areas of the water. This competition involves much chasing and several visual displays. Mating and egg-laying are equally entertaining but not as frequently seen.

How to Find White-Tailed Dragonflies

This is one of the most common dragonflies and it can be found all across North America. Go to the edges of ponds, lakes, or slow-moving rivers and look for dragonflies with a wide and flattened bluish-white tail and a large amber stripe across the middle of each wing. These are the males; they will be skimming back and forth above the water, every so often chasing another dragonfly. The females have a dark abdomen and are best found by watching the behavior of the males (see the following description).

What You Can Observe

DRAGONFLY BEHAVIOR

If you visit the edge of a pond or lake on a nice summer day, you are almost assured of seeing dragonflies darting about over the water. The activities of the various species may seem like random movements, but with a little knowledge of their behavior, you will be able to see recurring patterns in their interactions and even be able to follow the

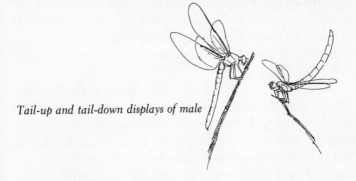

Tail-up and tail-down displays of male

fate of individual insects. Dragonflies are superb for this type of behavior-watching, for they are large insects, can be approached without disturbance, and have fascinating social behavior.

There are many species of dragonflies, and so few have been carefully studied that it would be a mistake to generalize about their behavior. Therefore, I have chosen one very common and easily recognized species and will describe its behavior in detail. After watching it and discerning its behavior, you can then look at other species and make some of your own behavioral comparisons.

Do your observing at midday, for this is when dragonflies are most actively engaged in social behavior. Most of the time you will see males, since the females generally only visit the water for short periods, during which they mate, lay eggs, and then leave. The males remain throughout the middle part of the day.

As it arrives at the pond, each male generally remains near a small section of the pond shore. Once it has chosen an area, it usually returns to the same spot each day it is at the pond. When there are no females present, there are three activities that you will see males engaged in: perching, chasing, and patrolling.

In his area, a male has one to three perches, which may be sticks or reeds protruding from the water, or rocks or logs just on shore. Perched on one of them, the male faces the center of the pond and follows the movements of other males near him.

Often, when another male enters the area, the "owner" will fly directly toward it and be involved in one of three types of chases. First, he may just chase the other male away; or, the two may face each other, fly up together a few yards into the air, and then chase back and forth across the area; or he may move to a position beneath the other and then quickly fly up, causing an audible clattering of wings as he contacts the other dragonfly. Then both fly way up into the air and off to one side.

The third activity, patrolling, consists of the male leaving his perch and flying about the area in a slow, meandering path, finally returning to one of his perches.

During all of these activities, a male is likely to compete for dominance over the area by signaling to other males. The signal for dominance is raising the abdomen to about seventy degrees from the horizontal and showing the whitened upper surface to another male. A signal of subordinance is the opposite, lowering the abdomen and hiding the white portion. During interactions, if you look closely, you will be able to see this raising and lowering of the abdomen. It happens quickly and may in some instances be a repeated flashing.

These three activities and two displays are used by the males to gain access to females. The areas that the males occupy are places where the females lay their eggs. As the females come to these areas, the males that are most dominant in them get the best chances of mating with the females.

In areas where there is a low density of males, there may

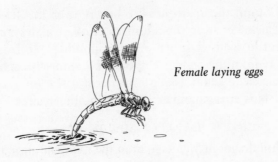

Female laying eggs

be only one male per area and he may be able to chase all other intruding males off. However, in areas where there are many males, there may be three or four males per area, but one of them will be dominant at any time over all the others. Keeping track of interactions and who is dominant is difficult without having individuals marked, but you can still get a feel for the area and the different males in and around it by watching the activities of a few males over the period of fifteen minutes or so.

BEHAVIOR OF FEMALES: MATING AND EGG-LAYING

When a female enters the area, generally a dominant male will approach her and grasp her thorax with his legs while hovering. Then, with the tip of his abdomen, he grips her behind her head. She then bends her abdomen around until it touches the second abdominal segment of the male, where his penis is, and sperm is transferred. This is all done in the air and takes only ten seconds or less. The male then releases the female and she starts to lay eggs rapidly by dipping the tip of her abdomen into the water and washing the eggs off as they are pushed out. She seems to favor laying eggs on glistening objects, such as floating leaved plants, or algae-covered rocks.

Egg-laying may last for over two minutes, and during

this time the male usually hovers over the female and protects her from the advances of other males that might want to mate with her. After ovipositing, the female flies away from the pond and the male resumes his activities in his area, possibly taking a pause of a few minutes before he starts engaging in activities with other males.

FEEDING BEHAVIOR OF ADULTS

Adult dragonflies emerge from the nymphal stage throughout the early part of the summer, rather than all at one particular time. For the first two to three weeks after emergence, they leave the water and feed in woods and fields. Dragonflies have long legs covered with hairs, which are used to catch other insects in the air. Dragonflies are magnificent fliers, and it is fun to watch them feed. Some use perches and watch for insects passing by; you can even see their heads turn as they follow their prey's flight path. When an insect is close enough, they dart out after it and bring it back to their perch to eat. Other dragonflies continue to fly back and forth over an area and dart after prey while in flight.

BEHAVIOR OF THE NYMPHS

It is difficult to observe dragonfly nymphs while they are still in ponds and streams, for the water is often murky and the nymphs dark. But usually, with a bucket, you can take

White-tailed nymph about to grab prey

a scoop of water and mud from a pond edge and find a nymph among the debris. They are one-half to three-quarters of an inch long, brownish, and have long legs and a strongly segmented abdomen. They breathe by drawing in water through the tips of their abdomens, extracting the oxygen, and then expelling it. When disturbed, they use the expelling as jet propulsion to move them quickly away. They feed with an amazing pair of pincers that normally fold up under their heads, but which can reach out quickly to catch prey slightly in front of them.

BLACK-WINGED DAMSELFLY

Relationships

The black-winged damselfly is a species (*Agrion maculatum*) of insect in the family Agrionidae, or broad-winged damselflies. The wings of broad-winged damselflies taper gradually to their base rather than being stalked as in other damselfly families. Damselflies and dragonflies form two large groups in the order Odonata. Damselflies have much thinner bodies than dragonflies, and for the most part they fold their wings over their backs when at rest, unlike dragonflies, which leave them out to the sides.

Black-winged damselfly displays. Left: cross-display; right: wing-spreading. Life size

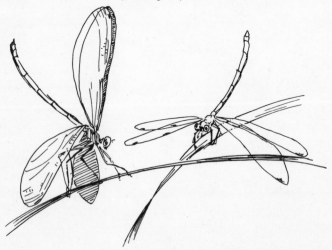

Black-winged damselfly habitat

Life Cycle

Black-winged damselflies overwinter as nymphs underwater. Mature nymphs crawl out of the water in late spring through summer on to reeds or nearby plant stems. While the nymph is securely holding on to the stem, a split forms in the back of the thorax and the adult crawls out. An hour or two later, the adult can fly and usually goes to wooded areas for feeding. Within a few days it returns to the water edge and mates. The female lays eggs on plant stems, usually just below the water surface, while the male that fertilized her waits nearby and keeps other male damselflies away. The eggs hatch in about a week, and the nymphs feed underwater on other insects and underwater animals. The maturing nymphs overwinter and leave the water the next spring to become adults.

Highlights of the Life Cycle

The most exciting stage of these insects' lives is when they are adults and come to the edge of streams to mate, for the males defend small territories and, along with the females, use a variety of displays to finally choose a partner and mate. These displays are easily seen and distinguished. It is also fun to watch egg-laying and feeding behavior.

How to Find Black-Winged Damselflies

Black-winged damselflies perform their mating activities along the sides of small, shady streams. When you are in these areas, look for damselflies with black wings and iridescent green bodies: these are the males. The females are all brown with a white dot at the front tip of each wing.

What You Can Observe

COURTSHIP AND TERRITORIAL BEHAVIOR

Along with their obvious beauty, black-winged damselflies also have some fascinating behavior that is easy to observe. The first thing to do in beginning to observe their behavior is to practice distinguishing males from females. Probably the easiest way to tell them apart is to look at the wings. Clearly visible on the tip of each of the female's wings, whether she is in flight or at rest, is a small white dot. If these are absent, the damselfly is probably a male.

Next, watch the movements of the males. From midmorning well into the afternoon, many males remain in, and defend, small areas of the pond or stream. These areas might be called territories and include two things: an area for a female to lay eggs in, and a small perching spot that overlooks that area. Suitable egg-laying areas are places where there is some vegetation or debris just at the surface of the water for the female to attach her eggs to.

Males with territories will do several things: they will either stay in one spot, or, if they fly away, they will soon return to that exact spot; they will chase other males away from the area; and they will do a display to approaching males called wing-spreading, in which they spread their wings all the way open and raise their abdomens. This display often makes the intruder move away.

The next thing to look for is interaction between males and females. When a female enters a male's territory, the male does another display called the cross-display. In this, the back wings are spread while the fore wings remain folded over the insect's back, and the abdomen is lifted way into the air. In response to this, the female may perch nearby or may fly off toward the shore. In either case, the male gets into a position where he is facing her while flying and does another display called the rapid-wing-flutter, in which he beats his wings at an obviously faster rate. At this point, the female will either fly off, do wing-spreading, or remain still on her perch. In either of the last two cases, the male will eventually try to mate with her.

MATING AND EGG-LAYING BEHAVIOR

The male first lands on the female's back, then takes the tip of his abdomen and fastens it just behind her head. He then pulls the tip of his abdomen up so that it touches his abdomen's second segment. The male's penis is actually on the second segment of his abdomen, but his sperm is produced at the tip of his abdomen. So what he has just done is transfer sperm to his penis. After this, the female bends the tip of her abdomen until it engages with the male's penis and she is fertilized. Copulation lasts about two minutes. Afterward, the male and then the female

Typical damselfly mating posture (Coenagrionidae *sp.*): *male on top with tip of abdomen attached just behind female's head*

return to the territory and the female starts to lay eggs for anywhere from a few minutes to up to an hour. Meanwhile, the male perches nearby and chases off any males that approach too close to her. At the same time, the territorial male may perform the cross-display to other females but will not continue with the other mating displays until the first female has left.

Males that could not defend a territory or do not have one for other reasons may still mate with females. Their ritual is similar to the territorial male's, except that they omit the cross-display and start with the rapid-wing-flutter. Females mating with nonterritorial males then need to find an egg-laying site, but most of these are guarded by territorial males, and in order for these females to lay eggs there, they must first mate with the territorial male, or find a less favorable egg-laying site that is not guarded.

DAMSELFLY NYMPHS

To observe damselfly nymphs, remove them from water and put them in a clear container. They are slender, about a quarter to three-quarters of an inch long, and have long legs. At the tips of their abdomens are three featherlike projections, which are gills through which the nymph extracts oxygen from the water. This is different from the dragonfly nymphs, which take in water through their abdomens and then expel it, enabling them to move using

Damselfly nymph

jet propulsion. Damselfly nymphs do not have this ability, but rather move by flipping the gills from side to side, much as a fish does with its tail. This behavioral difference in movement is a way to distinguish the two types of nymphs.

FEEDING BEHAVIOR

Black-winged damselflies do not spend all of their time along streams holding territories, displaying, mating, and egg-laying. A great deal of time is spent deep in the woods, feeding. Sometimes damselflies position themselves on a branch near a patch of sunlight. As small insects fly into the light, they are lit up and easily seen against the dark background of the rest of the trees. The damselflies fly out, catch the insects in mid-flight, and return to their perch to eat them. Insects are caught in the legs of the damselfly, which are long and covered with hairs.

LARGE MILKWEED BUG

Relationships

The large milkweed bug is a species (*Oncopeltus fasciatus*) of insect in the family Lygaeidae, or seed bugs, which, in turn, is in the order Hemiptera, or true bugs. The common name of the family refers to the fact that many members feed on developing seeds. Another member of this family, the small milkweed bug (*Lygaeus kalmii*), resembles the large milkweed bug and also feeds on milkweed.

Life Cycle

Large milkweed bugs overwinter in the adult stage. In early summer some adults migrate to northern areas. Adults mate in early summer, and females lay eggs in clusters of about fifteen per laying on the undersides of milkweed leaves. The eggs hatch in three to six days and the nymphs feed on milkweed. They undergo five molts during a period of

Milkweed bug habitat

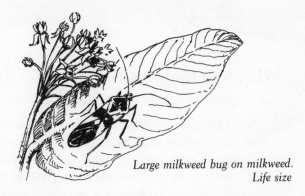

Large milkweed bug on milkweed.
Life size

thirty to forty days. As adults, they are able to fly. Adults
may live thirty to forty days, and there may be two or more
broods. Many adults in northern areas migrate south in
fall; the nymphs and adults that remain in northern areas
die in the freezing temperatures.

Highlights of the Life Cycle

All stages of this bug's life can be seen on milkweed plants,
and this provides a marvelous opportunity to watch gradual
metamorphosis in detail. You can also watch adult mating
and egg-laying behavior. Another feature of this insect is
its habit of migration; this is hard to observe but fun to
know about.

How to Find Large Milkweed Bugs

Look under the leaves and among the flowers of milkweed
plants for a red and black bug about a half inch long. The
black will be at both ends and in a band across the middle
of the insect. If you are in the northern half of the conti-
nent, you may not find the bug until the middle of July,
for the adults have to migrate. Approach the bugs slowly,

since they have the habit of dropping into the grass when disturbed.

What You Can Observe

THE VARIETY OF RED AND BLACK INSECTS ON MILKWEED

There are a number of other insects that feed on milkweed that you may confuse with the large milkweed bug. First, there are two or three common species of beetles that are red with black spots on their backs. They can easily be recognized by seeing that there is a straight line down their backs where their wing covers meet and touch without overlapping. One common species has just four dots on its back; another, six spots on its back; another has larger blotches. The larvae of these beetles burrow into the stems and roots of milkweed.

The large and small milkweed bugs both have triangles on their backs where their wings fold over each other. The small milkweed bug has an X of red on its back, whereas the large milkweed bug has two bands of red across its back.

It is interesting that these insects on milkweed all have red and black coloration. One explanation is that there is a chemical in milkweed that is harmful to many animals. Insects that have evolved immunity to this chemical and feed on milkweed may contain the poisonous element, and because of this gain some protection from predators. But

Large milkweed bug nymphs on leaf, milkweed beetle on stem

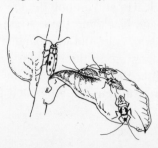

they must also warn predators that they may be dangerous, and the color patterns of black and red may be such a warning. Monarch butterflies contain the poisonous chemical, for their larvae feed on milkweed and they are also black and red.

BEHAVIOR OF THE NYMPHS

The majority of adult large milkweed bugs will be seen in late summer and fall feeding on the pods and seeds of milkweed. Since these are true bugs, they have gradual metamorphosis, developing from an egg, through five nymphal stages, to the adult stage. The eggs are laid on the undersides of the milkweed leaves and are arranged in loose clusters, each containing about fifteen eggs. The nymphs feed on the flowers or pods of milkweed and mature in about forty days. The nymphs often feed in large groups clustered on a single milkweed pod, and unless you approach them carefully they will all drop to the ground. Like many other types of bugs, they may remain motionless for several minutes on the ground before climbing back up. During this motionless state they appear dead, and this may give them some protection from predators that are more likely to eat a moving insect.

MATING

The adults may often be seen mating. To do this the male crawls onto the back of the female, connects his abdomen to hers, then loosens his hold and moves around until the two are facing in opposite directions and connected end to end. The two may move around in this posture, the female generally dragging the male. The length of time the two spend in copulation depends largely on the temperature. In warm weather they may remain attached for from ten to thirty minutes, but in the colder fall weather they may remain together for more than a day.

MIGRATION

When the large milkweed bug was first closely studied (1932), the investigator could not find any evidence of the insects in winter. He knew that farther south in Missouri there were some adults active through winter, but when he caged various adults and nymphs outdoors in his area of Iowa, they all died over the winter. He also observed that just south of his area the adult bugs always appeared a week earlier than just slightly north of his area.

Recent studies have at least partially solved this mystery. The large milkweed bug does not overwinter in the northern parts of its range, but migrates south to more favorable conditions, where it lives through the winter as an adult. However, not all of the bugs migrate; only about a quarter of the bugs tend to make extended flights. Throughout most of the summer, the insects do not migrate, but as the daylight period shortens, the days get colder, and the population of bugs increases, the reproduction activities of adults are delayed, creating a greater tendency in them to fly.

CICADAS

Relationships

Cicadas are a family (Cicadidae) of insects in the order Homoptera, or bugs. As nymphs, cicadas remain underground and feed on plant juices from roots, sometimes not emerging for ten to twenty years. Male cicadas use long, continuous buzzes as part of their mating behavior. There are many species, each making a slightly different buzz.

Life Cycle

Cicadas overwinter as nymphs underground. The nymphs that are mature in early summer tunnel up through the

Cicada habitat

The spilt-open empty nymphal skin, from which the adult cicada emerged. Life size

ground and crawl onto the trunks of trees or some other surface that will support them. After latching onto the surface with their claws, the mature winged cicada emerges from a slit along the back of the nymph's exoskeleton. The adults fly into trees and feed on sap from twigs. The males make sounds to attract females for mating, and the females lay eggs in slits that they make in tree twigs. The young nymphs hatch, fall to the ground, and burrow underneath to feed on sap from plant roots. In the case of the seventeen-year cicada, the nymph stays underground seventeen years, but there are many other common species in which the nymphs complete their development in one to three years.

Highlights of the Life Cycle

One of the real highlights in the lives of Cicadas is the beautiful buzzing call given by the males on hot summer days. Each species has a different type of call and you may be able to distinguish among them. It is also exciting to go looking for the nymphal skins attached to tree trunks where the adults emerged from them. Below these, in the ground, you may also find holes where the nymphs burrowed out of the soil.

How to Find Cicadas

Cicadas are generally in areas of tall shade trees, and are especially common in the suburbs. First, listen for the call, which is a long, continuous buzz that swells in intensity and loudness and then dies off near the end. Where you hear the call, check tree trunks for the shed nymphal skins, and where you find these, check the earth beneath for nymphal emergence holes. They are about the size of your little finger and have no signs of excavated earth around them.

What You Can Observe

CICADA SOUNDS

Just about everybody confuses cicadas with locusts, because cicadas are commonly called seventeen-year locusts. This is a misnomer, for locusts are related to grasshoppers, katydids, and crickets, and are in the order Orthoptera. Cicadas are actually bugs, and are in the order Homoptera, along with leafhoppers and aphids. Locusts and grasshoppers also make sounds in summer and this adds to the confusion. (See Crickets and Grasshoppers, fall section.)

The sound of cicadas is made in an entirely different way from the sounds of other insects. The grasshoppers, locusts, and crickets make sounds by rubbing together two body parts, such as leg against wing, or wing against wing. This is not the case with cicadas. In the last segment of the thorax, there are two hollow cavities covered on one side with membranes that act like drum heads. Attached to these membranes are muscles that cause the membranes to vibrate. Most of the mass of the large abdomen of the adult cicada is empty, a large hollow chamber, and this may help amplify the sounds produced in the thorax.

Adult cicada on oak twig

The sounds of cicadas are made by the males and have the effect of attracting other males to the same spot and stimulating them to sing as well. This can result in a large number of males all gathered in the same trees and all calling. This chorus of males in turn attracts females from as far as they can hear the sound, and when they arrive they choose males and mate. Each species of cicada, and there are over seventy-five in eastern North America, has a distinct call. The calls are used to help species distinguish each other and find proper mates. This is further helped by each species singing at its own time of the day and in its own particular habitat. Some species sing in the middle of the day and others are more attuned to singing in the evening.

LOOKING FOR NYMPHAL SKINS

Once you hear cicadas singing, you can start looking for their shed nymphal skins attached to the trunks of trees. The nymphs usually crawl up to the treetops to cast their skin, but may also do it lower on the trunk, where we have the chance to find them. Look up and down the tree trunk for the light brown castoff skin of an insect, about an inch long and a little less than a half inch in diameter. It will be attached to the bark by the sharp tips of four legs, and its front two legs will be modified into claws, much like those on the front legs of an ambush bug (see fall section). The thorax is split open along the back where the adult crawled out. Inside the cast-off skin are whitish tubes that were the spiracles, or breathing tubes, which penetrated the nymphal skin, allowing the insect to breathe.

NYMPH BURROWS

While maturing, these nymphs lived in tunnels underground, where they fed on the sap from tree roots. Just before emerging, the nymphs burrow up until they are just short of the surface and make a small cavity in the earth

Right: *young nymph in burrow*. Left: *emerging full-grown nymph*

that may be as much as several inches long. They wait there until evening and then dig through the remaining earth and climb up the nearest trees. It is fun to look for their emergence holes. They are about five-eighths of an inch in diameter, in hard-packed earth near the base of a tree. There is no evidence of excavated earth near them because the nymph has an ingenious way of loosening earth in front of it and packing it on the sides and rear of the burrow. To do this it uses its highly modified front feet, which serve the functions of a pick, a rake, and a tamper all in one.

The nymph stage of cicadas is among the longest of any of our insects, ranging from two to twenty years underground, depending on the species and possibly the food conditions as well. Periodical cicadas have life cycles of either seventeen or thirteen years. In different parts of the country, different broods of periodical cicadas exist, so that every year, somewhere, there is an adult stage emerging. The periodical cicadas are so often written about that most people are unaware that our most common cicadas have life cycles of only a few years.

EGG-LAYING SITES AND NYMPHAL DEVELOPMENT

Once the cicadas have mated, the females lay eggs in the young twigs of trees by making slits in the wood and depositing the eggs. The mechanical action of making the slits often kills the twig. This is particularly evident on oaks and hickories in late summer. Where females have been active, you see little bunches of brown leaves among the greenery of the rest of the trees. In some trees, the twigs actually break off in the wind. Cicadas do not seem to be particular about what trees they lay their eggs on, except that they usually avoid conifers.

After hatching from eggs on the tree twigs, the nymphs

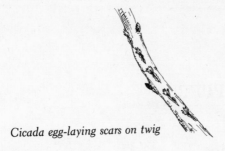

Cicada egg-laying scars on twig

fall to the ground and immediately begin to dig down toward the roots, where they will feed. You will not see this, for the nymphs are too small and are soon underground. In fact, after seeing the twigs killed by the female's egg-laying, there are no more signs of cicadas until singing starts again the next year.

LEAFHOPPERS

Relationships

Leafhoppers are a family (Cicadellidae) of insects in the order Homoptera, or bugs. They are small, elongate bugs that feed on the juices of plant stems and hop about from plant to plant. They are one of the most common insects in any grassy area, and many have brilliant colors on their front wings, which are folded over their backs when at rest. They may be confused with treehoppers, which are larger and feed on woody stems, or spittlebugs, which are more oval and have spittle around them in the nymphal stages.

Life Cycle

Leafhoppers overwinter in the adult stage. They become active in late spring and summer, feeding on juices from

Leafhopper adult on grass blade. Life size

Leafhopper habitat

plant stems, mating, and laying eggs in the stems and leaves of grasses and wildflowers. The nymphs hatch in one to two weeks. During the next month they undergo five molts and mature into adults. Depending on the latitude in which they live, there may be from one to four or more broods, the larger number of broods occurring in the warmer areas. Adults of the last brood crawl down to the base of grasses and overwinter.

Highlights of the Life Cycle

The nymph and adult stages are the only ones easily seen. Of these, the adults are the most entertaining to observe, as they feed, mate, hop about, and lay eggs. Many species are brilliantly colored and a joy to see close-up.

How to Find Leafhoppers

Leafhoppers are so common in any area of tall grasses or weeds that it is hard not to find them. Simply brush your hand over the grasses or wildflowers and you will see the insects hop out in all directions. You can also sit down among grasses and put a piece of paper on your lap or in the grass and watch them land on it as you disturb the area around it. They are elongate insects about a quarter inch long. Their wings are folded over their backs in tentlike fashion. They are best distinguished from the similar spittlebugs by the presence of spines along the sides of each hind leg. Spittlebugs have only one or two spines on their back legs.

What You Can Observe

LEAFHOPPER ACTIVITY

In the earliest spring, even before the leaves come out, you may see leafhoppers flying about, especially in the vicinity of apple trees. These are adults that overwintered and are now locating mates and mating. When the apple leaves emerge, the females lay their eggs on the midribs of the leaves, and the young nymphs feed on the leaves.

Leaves that leafhoppers have fed on have small dots and are sometimes curled or browned at the edges. If you watch one of the larger leafhoppers while it feeds, you may see why they have been called sharpshooters: as they drink the sap, they expel tiny droplets from the tips of their abdomens, and these are so forcibly expelled that it seems as if they are "shot." A red-banded leafhopper was reported by one observer to have expelled one droplet per second for up to two minutes.

As you approach leafhoppers, especially the nymph stages, they will try to run away from you. Here you will notice a curious event: they run sideways. As they get older, they have more of a tendency to hop, and as adults they will run, hop, or fly. The flight of the adults can be amazing, for some species at the edge of their northern

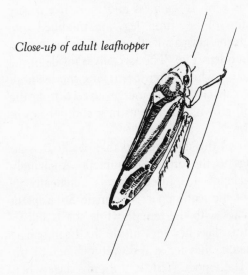

Close-up of adult leafhopper

range have been observed to fly about a hundred and fifty miles north in the spring to take advantage of food plants, and then in fall to fly back to avoid the colder temperatures for overwintering.

LEAFHOPPER PARASITES

Less readily observable, but still common, are three groups of parasites associated with the leafhoppers. One is a tiny fly in the family Pipinculidae. These flies are superb at hovering, and they hover in the areas where leafhoppers feed. When a leafhopper is seen, the female flies down, picks it up, and returns to hovering. While hovering, it lays an egg on the leafhopper and then drops the insect. The larva grows inside the leafhopper until it completely fills the body cavity of it and then breaks out and pupates in the ground.

Another parasite is a small wasp in the family Dryinidae. The female of this wasp is wingless and looks like a very small ant. The tips of her front legs are modified into pincers with which she grasps a leafhopper, lays her egg on it, or in it, and then lets it go. The larva lives inside of the leafhopper for a short while, often destroying the leafhopper's reproductive capacity, then emerges to feed from the outside. When full grown, it leaves its host and pupates in a tough silk cocoon.

The last is perhaps the most remarkable of all. It is an insect that belongs to the order Strepsiptera. Amazingly enough, the female of this insect spends practically its entire life inside of the leafhopper. The male Strepsiptera flies about and mates with the female while she is still in the host. She in turn lays her eggs inside the Leafhopper, and when these hatch, they leave and wait to attach themselves to another leafhopper. This is the only stage when the females are not in their host. Strepsiptera habits are similar to those of the blister beetles that prey on solitary bees (see Solitary Bees, spring section).

TREEHOPPERS

Relationships

Treehoppers are a family (Membracidae) of insects in the order Homoptera, or bugs. They are small, common insects that feed on the sap from woody plants. Like leafhoppers and spittlebugs, the adults can hop long distances (a yard or more). A distinctive feature of this family is that a segment of their exoskeleton on the back of the thorax, the

Treehopper habitat

Two-spotted treehopper and egg masses on bittersweet. Life size

pronotum, is often greatly enlarged into fantastic shapes that resemble thorns, twigs, and buds. These probably help to camouflage the insects.

Life Cycle

Treehoppers overwinter as eggs laid on the twigs of woody plants. They hatch in early summer and start to feed on sap from plants. Nymphs undergo five molts in about six weeks and then are adults and can fly. The adults mate and lay eggs on woody twigs. There are one or two generations a year, the greater number of generations occurring in the warmer climates. The eggs produced by the last brood overwinter.

Highlights of the Life Cycle

The adults are beautiful to see because of their unusual shapes. Wherever you find the adults, you can also usually look for the eggs. The two species discussed here exemplify two different methods of protecting the eggs through winter.

How to Find Treehoppers

Two of the more common species of treehoppers are the two-spotted treehopper and the buffalo treehopper. The adults of these species are quite plant-specific when feeding, so the easiest way to find the insect is to look for the plant first. The two-spotted treehopper can be found on the twigs of locust, bittersweet, hoptree, and butternut. They look like little black thorns a quarter inch high. Look also for tiny masses of white material about an eighth inch long — these are the insects' egg clusters. It is easiest to spot the egg coverings and then to look for the adults. The buffalo treehopper is best found on clover and aster in summer. Look on the stems of these plants for a green, stocky bug with a flattened portion starting at its head and projecting over its back. It is also a quarter inch long.

What You Can Observe

THE TWO-SPOTTED TREEHOPPER

Although these bugs are called tree "hoppers," they do not hop from branch to branch. When disturbed, they usually either crawl around to the opposite side of a branch, or else, if adults, hop into the air and fly a short distance away. Hopping is used mainly as a way of starting flight.

The two-spotted treehopper is often easier to find than the buffalo treehopper because of the conspicuous white covering that the female builds over the eggs. These little patches of white, sticky froth can be found at almost any time of year, for they persist even after the young have hatched. Look for them, especially in late summer and early fall, on the tips of the branches of the bug's host plants: locust, bittersweet, hoptree, and butternut. To lay the eggs, the female makes a slit in the bark with her ovipositor and deposits the eggs. Then, by rocking her abdomen from side to side, she deposits the white marshmallowlike substance over the slit. Usually there are several egg clusters laid along the same area of the twig. Once you have found the egg cases, look for the adults. They are shaped like small thorns, and are dark brown or black and have two light spots on the pronotum.

The eggs of this treehopper are laid mostly in August and September, and become particularly obvious in winter when the leaves fall from their host plants. The insect overwinters in the egg stage and hatches in early spring.

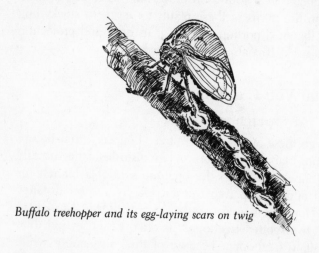

Buffalo treehopper and its egg-laying scars on twig

Both the nymphs and the adults feed on the sap of the host plants. The young nymphs stick their mouths through the bark of young shoots and suck the sap. They stay with the newest growth, for older twigs have thicker bark that is hard for the small insects to poke through. The nymphs have no wings and so cannot travel far from where they hatched. But once they become adults, they can fly and, in the case of this species, often move to more succulent plants, such as joe-pye weed and daisy, to feed on in the summer. Then they move back to their host tree in late summer and fall to lay eggs.

THE BUFFALO TREEHOPPER

The buffalo treehopper is harder to spot since it is green and blends in with its food plants. Also, it does not cover its eggs with anything but just lays them in slits in the bark. The life cycle is essentially the same as that of the two-spotted treehopper, except that the young nymphs leave their place of hatching — the twigs of elm or apple trees — drop to the ground, and feed on clovers and asters. Even the adults may be found on these plants in midsummer, but when they are ready to lay their eggs in late summer, they move to the twigs of elm and apple trees. The egg slits of this species are distinctive and you may come across them as you examine twigs in winter and fall. They are two curved slits, about a quarter of an inch long, resembling a set of parentheses. There are usually a number of pairs of slits in succession along the twigs.

The nymphs go through five molts before becoming winged adults, and during this time they are hard to find, since they nestle into bark crevices and leaf axils.

Various species of treehopper adults

APHIDS

Relationships

Aphids are a family (Aphidae) of insects in the order Homoptera, or bugs. They are minute insects that feed on plant sap and reproduce very quickly. For most species, there are both winged and wingless generations. Aphids have the ability to give off excess plant juices in a sweetened droplet called honeydew, which many other insects come to feed on. Sometimes relationships of mutual benefit develop between the aphids and the insects that feed on their honeydew.

Aphid habitat

Aphids on milkweed. Life size

Life Cycle

Aphids overwinter as eggs placed on or near their spring food plants. These hatch in spring and the nymphs are all wingless females. They in turn give birth to live young, rather than eggs, and these also develop into wingless females. There may be as many as thirteen generations in a single summer, all produced by the females without males. This is called parthenogenesis, or virgin birth. Some of these females at various times in the summer will give birth to winged females, which will migrate to other plants and start new colonies of wingless females. Anytime from mid-summer to late fall, depending on the species, one of the generations of females gives birth to both males and females. These then mate and the fertile females lay eggs. It is these eggs that overwinter and hatch into female nymphs in the spring.

Highlights of the Life Cycle

Aphids can be easily found through most of the summer. You will be able to see various sizes of nymphs, winged and wingless adults, and you may even see them giving birth. Along with all of these features is the likelihood that you will find the aphids attended by ants. The interrelationship between these two insects is also intriguing to observe.

How to Find Aphids

Aphids are about one-eighth inch long when full grown and are almost always found in large colonies of fifty to a hundred or more. They come in many colors and are most often found in weedy areas, clustered at the tips of young plants, which provide the easiest source of sap for them to feed on. Look in early summer when meadows and fields are still lush. In late summer when things have dried out considerably, it is harder to find aphids.

What You Can Observe

APHID LIFE STAGES

Not all the aphids in a cluster are the same size. Some are very small, no bigger than a pinhead. These small ones have recently been born and are gradually developing into adults. Aphids are born alive in summer rather than as eggs in spring. The largest aphids in the cluster are adults, and

Aphid giving birth to live young

you may even see one giving birth, since this is a common occurrence. Another thing you may see is light brown shells in the area of the aphids. These are the skins shed by the nymphs as they develop. Some adult aphids may have small sets of clear wings, but the majority will be wingless (see Life Cycle).

RELATION TO FOOD PLANTS

Ninety percent of all aphids live only on one species of plant and are specific to that plant. They feed on the growing parts of the plants by inserting their mouthparts into the stem and sucking out the sap. The growing part of the plant is usually the only part with sap rich enough to sustain Aphid growth. Therefore, when the plant stops growing, the Aphids must stop as well. Aphids solve this problem in two ways. First, they may simply stop reproducing and go into a summer hibernation, called aestivation, until the plant starts growing again. This is especially common in many Aphids that use woody shrubs or trees as hosts, for these plants typically have two periods of growth, one in spring and one in late summer or fall. Or, they may move to a new host still in an active phase of growth.

APHIDS AND ANTS

As aphids feed, they take in plant sap, extract the nutrients from it, and excrete the excess fluid in a small drop of what is commonly called honeydew, for it contains sugars in a liquid form. Aphids normally develop just a small drop of this fluid at the tips of their abdomens, and then give a little flick of their abdomens to shake the drop loose. These drops of honeydew accumulate on the leaves and vegetation below the aphids and, in turn, are often fed upon by other insects, such as butterflies, flies, and bees. Also, there is a black mold that commonly grows on this residue and may

Ants tending aphids

make the leaves of a tree or plant appear to be covered with sticky soot.

You will frequently find ants crawling over the aphids. This is not just a chance occurrence but a regular event. If you continue to watch several ants on the aphids, you are likely to see one ant touch the abdomen of an aphid and then eat a drop of honeydew excreted by the aphid. Many species of ants seem to favor honeydew as a food source and develop regular associations with aphids feeding near the ants' nest. Members of the nest repeatedly visit the aphids and collect honeydew. They have also frequently been observed to attack or be aggressive toward most of the common predators of the aphids, such as the larvae of ladybird beetles, syrphid flies, or lacewings. This seems to be an obvious advantage to the aphids, since it protects them, and is also an obvious advantage to the ants, since they get an undisturbed source of honeydew from the aphids. It is of mutual benefit. The ants' presence seems to have an effect on the behavior of the aphids. Aphids "milked" by ants tend to hold on to their excess honeydew longer, making larger droplets available to the ants. Without ants, the aphids hold on to less honeydew before shaking it off. Another interesting change is that aphids tended by ants reproduce more rapidly and produce proportionately more honeydew.

APHIDS AND THEIR PREDATORS

There are three insect larvae and two insect adults that commonly feed on aphids. The larvae are those of ladybird beetles, syrphid flies, and green lacewings. Syrphid fly larvae are sluglike, about a quarter inch long, have no prominent legs, and are usually light green. Lacewing larvae are similar in size and shape but have three obvious pairs of legs and long, sickle-shaped pincers at the front of their heads. Ladybird beetle larvae are slightly smaller, dark colored, have three pairs of obvious legs, but no pincers on their heads.

The two adults that feed on aphids are very different from each other. The ladybird beetle is hemispherical, about one-eighth inch long, and often brightly colored red with black dots. The green lacewing is about three-quarters of an inch long, with a thin, green body, long antennae, and large membranous wings folded tentlike over its back.

Life of the Ladybird Beetle: Ladybird beetles, sometimes called ladybugs, are familiar to us all, at least in their adult stage. The fertilized female seeks out a colony of aphids,

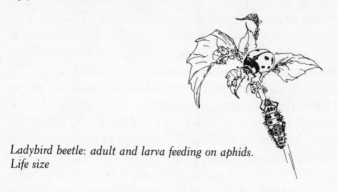

Ladybird beetle: adult and larva feeding on aphids.
Life size

not only to feed herself on but also as a place to lay her
eggs. She lays small groups of orange eggs on the undersides
of leaves, usually within five inches of the aphid colony.
The young hatch in about eight days and search the plant
for aphids. Even though the aphids are larger than the
beetle larvae at this stage, the larvae have no trouble eating
the aphids. The larvae go through three molts in about two
weeks. At the end of this period, they attach the tips of
their abdomens to the top of a leaf or stem near the colony
and shed their skins once more before pupating. They
emerge as adults from the pupae in another week. The
complete life cycle takes about a month. One nice thing
about this beetle is that you have a unique chance to see
all stages of its life by simply looking carefully among the
leaves surrounding the colony of aphids. The larvae, when
full grown, are only about a half inch long. The two-
spotted ladybird beetle is one of the most common, and in
its adult stage it has an orange back with a dark spot on
each side.

Life of the Green Lacewing: The lacewing is an insect
that, as an adult, is often attracted to lights, so is often
found inside your house in the morning, resting on a wall

or ceiling. It is green with lacelike wings, golden eyes, and long, hairlike antennae. It can also be found feeding on the aphids as an adult, and it lays its eggs on the leaves or stems near an aphid colony. Its eggs are easily distinguished

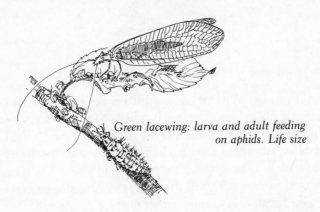

Green lacewing: larva and adult feeding on aphids. Life size

from those of the ladybird beetle, since each is placed at the tip of a thin, hairlike stalk about a quarter to a half inch long, which is created by the lacewing. These are a beautiful sight to see and sometimes take on the appearance of a mold on the leaves. The young hatch in about five days, seek out aphids, and feed and grow for about the next fifteen days. They use their long pincers to catch the aphids and then actually suck out the juices of the aphid through the tips of the pincers. When mature, they spin a small cocoon either on a leaf or on the ground, shed their skin once more in the cocoon, and pupate. From egg to adult takes about a month, depending on the availability of food and the air temperature.

Life of the Syrphid Fly: While watching aphid colonies, you may spot a syrphid fly buzzing about the area. It may dart down, touch the aphids briefly, and then fly up again

Syrphid fly: adult hovering and larva feeding on aphids. Life size

to hover nearby. What it has most likely done is deposit an egg among the aphids. The egg hatches and the larva crawls about, feeding on the aphids. It has a characteristic method of grabbing an aphid larva and lifting it up into the air as it feeds on it. The syrphid fly remains in the larval state for two to three weeks and then pupates near the ground. (See also Syrphid Fly, summer section.)

Wasp parasites: Look for one or more aphids that are swollen up larger than the others and that may have turned brown. These probably are dead and very likely have wasp parasites in them. There are several species of braconid wasps that lay eggs inside the bodies of aphids. The wasp larva lives inside the aphid; when mature, it cuts a hole

Braconid wasp laying eggs in aphid; below: parasitized aphid showing exit hole

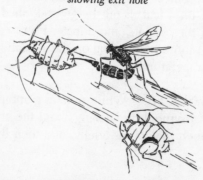

through the bottom of the aphid and attaches the aphid to the leaf with silk. It then goes back into the aphid and makes a cocoon of silk, in which it pupates. The adult wasp crawls out through a hole cut through the top of the aphid's abdomen. Another species builds a cocoon underneath the aphid and the adult emerges through a hole in the cocoon. There is even a smaller wasp that looks for the parasitized aphids and sticks its ovipositor through the aphid and into the parasite. Its young live on the body of the parasite.

TIGER BEETLES

Relationships

Tiger beetles are a family (Cicindelidae) of insects in the order Coleoptera, or beetles. They are mostly one-half to three-quarters of an inch long and are predacious as both larvae and adults, feeding on other insects. They have long,

Tiger beetle adult. Life size

slender legs as adults and are extremely quick when either running or flying. The young live in tunnels in the ground and catch insects passing by; the adults have long legs and strong wings, and they feed on other insects by pursuing them on the ground or in the air.

Life Cycle

Tiger beetles with a one-year life cycle overwinter as larvae in vertical tunnels in the ground. In spring they feed on insects passing by the tunnels. They pupate in the tunnels in summer and then emerge as adults. Females, after mating, lay eggs singly, just under the soil. The larvae hatch

Tiger beetle habit

in about two weeks and each immediately begins digging a tunnel, which it stays in except when reaching out to catch prey. They undergo three molts and overwinter as larvae.

In species with a two- or three-year cycle, eggs are laid in summer and the larvae overwinter, then are active for a summer or two. At the end of their first or second summer, they pupate and emerge as adults. The adults in fall dig a small burrow in which they overwinter. The next summer they mate and lay eggs.

Highlights of the Life Cycle

The life of the larvae is fascinating, but the larvae are secretive and not easy to observe, even though it is easy to find their tunnels. The adults are extremely common, but since they are so quick they are a real challenge to get close to. Some of them have beautiful iridescent colors, and these are well worth the effort to try to see.

How to Find Tiger Beetles

Our most common tiger beetles remain in dry areas bare of vegetation, such as dirt roads or paths. They are only active when it is bright and sunny. To spot them, walk about in their habitat and look six to ten feet in front of you for dark insects about three-quarters of an inch long, flying up from the ground and landing about fifteen feet farther away. You can get closer to one only by watching it land and then sneaking up very slowly. When airborne, they look a little like large houseflies, for they are very swift. They are extremely wary insects.

What You Can Observe

THE ADULT BEETLES

Whenever you come across a patch of bare earth or sand that is in full sun, you should think "tiger beetles." You have probably never heard of tiger beetles, let alone seen them, for they are rarely written about in popular books, and they are hard to see. But they *are* common, and once you learn some clues to their whereabouts, you will be able to find them easily and enjoy observing their behavior throughout the summer.

Three feet is about as close as you can get to these beetles, but from this distance you can see their important features. They have long, delicate legs that hold their bodies well off the ground and help them to run quickly. Many species have dark-colored backs with a few lines or dots on the wing covers. Often the underparts are iridescent, but depending on the light, you may or may not see this. Some species have brilliant iridescence on top as well. One species in particular, the six-spotted green tiger beetle, is a brilliant iridescent green and one of the most beautiful native beetles around. This species generally is a little slower than other ones and can be more readily approached.

The beetles' quickness and alertness certainly help them avoid predators, but also help them gather food, since they prey on other insects, catching them by either running or flying. They are especially fond of ants, but also eat a wide variety of other insects and small animals, including bugs, caterpillars, flies, worms, aphids, and even small fiddler crabs in areas by the shore. The beetles' mode of eating is usually to seize the prey with its mouth and then bang the prey against the ground several times until it is dead. Then they suck the juices out and chew parts of the body shell. It is both informative and entertaining to spend time watching these beetles feed.

THE LARVAE AND THEIR BURROWS

In these same sunny, dry areas, you can also observe some of the larvae's habits. The adult females, after mating, deposit their eggs one at a time just under the surface of the ground. After about two weeks, the egg hatches and the tiny larva digs a tunnel an inch or two into the ground. It then waits in the opening of the tunnel until a prey insect comes by. It then rushes out, grabs it, pulls it down into the burrow, and eats it. When the larva is ready to molt, it seals off the burrow and digs deeper down, sheds its skin, and then emerges to the top of the tunnel again.

In the barren areas where the larvae live, there are other

Larva of tiger beetle in burrow

types of small holes in the ground made by solitary bees, wasps, and ants, but you can distinguish among them. Bee and wasp burrows are generally the diameter of a pencil or larger and there is usually excavated earth to one side of the burrow entrance. Ant-burrow holes are about the size of a pencil lead and the excavated earth is generally placed in an even mound all around the hole. Tiger beetle larval burrows are the size of a pencil when the larvae are full grown, and smaller when they are younger. There are no signs of excavated earth around the hole; it is perfectly level with the ground. The hole entrance is usually slightly rounded at its lip, and for a radius of about one-half inch around the burrow entrance, the ground is usually kept clear of debris.

The burrows are always in open ground but may be on slopes up to forty-five degrees. They start out only a few inches deep but are made deeper depending on the age and species of larva, the soil conditions, and the season. In most native species whose life histories are known, the larvae spend at least one winter underground before maturing to adults. These larvae deepen their burrows up to four feet in winter to avoid the cold.

In the larval burrows, you may be lucky enough to see a larva poised at the entrance. The region behind its head is flattened and, when positioned correctly, it just closes up the entrance to the burrow and acts like a camouflage. This is not a common sight, though, for the larvae are very sensitive to vibrations in the earth and will retreat into their burrows when they feel you walking nearby.

FIREFLIES

Relationships

Fireflies are a family (Lampyridae) of insects in the order Coleoptera, or beetles. They are best known for their ability to make controlled flashes of light, but in fact not all species can make light. The species that can are nocturnal and use the flashes to attract mates. Firefly larvae are also largely nocturnal and live in marshy areas. Very little is known about them. Fireflies are carnivorous (eat other animals) in both larval and adult stages.

Life Cycle

Fireflies overwinter as larvae buried in the soil. They emerge in spring and continue to feed in swampy areas. In early summer they are ready to pupate, and make a small earthen cell in which to do this. Pupation lasts for about two and a half weeks. Then the adults emerge, fly about

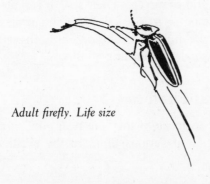

Adult firefly. Life size

Adult firefly habitat

at night giving light signals, and mate, and the females lay eggs for a couple of days thereafter. The eggs are laid on or just under the soil, and they may take up to four weeks to hatch. When the larvae hatch, they begin feeding and continue until fall, when they burrow underground to overwinter.

Highlights of the Life Cycle

Studying the system of flashes of fireflies is undoubtedly one of the most magical and engaging activities to be enjoyed in the world of insects. Their system is quite complex, and understanding it gives one a much greater appreciation for the workings of evolution, as well as just a greater sense of wonder about life.

How to Find Fireflies

Fireflies are easy to locate if they are near you; simply go out or look outside at various times in the evening and watch for small, flashing lights. Some of the best places to find them are over meadows or lawns, and at the edges of woods or streams. It is important to check at different times of the night, for each species is active in its own specific time.

What You Can Observe

THE SIGNAL SYSTEM

Most of the fireflies you see are males, since in the firefly communication system the males signal while flying and the females signal from perches on or near the ground. In some species the females are actually wingless. As the male flies over areas where there are likely to be females, he gives off the flash pattern typical of his species. If a female of the same species sees this, she responds from her perch with her signal. When the male in turn sees the female's flash, he continues to signal and moves closer to her. Through a series of signals and responses, the male finds the female and mates with her. The female generally stops signaling after this, although in one species it has been repeatedly observed that after being mated with her own species, she signals to males of other species, and when they come near she captures them and eats them.

As you watch fireflies, you can look for the signal patterns of different species. The patterns may differ in any of several ways: the duration of the signal, the interval between signals, the number of flashes in a complete signal, the distance the insect flies between signals, the color of the flashes, and whether the signal is composed of one or many flashes. One of my favorite pastimes is looking for new species' signals on summer evenings.

Two other ways that fireflies isolate their own species involve time of day and habitat. Each species tends to give

Firefly larva

signals for a fixed period of the day. This is always around dusk, some starting just before sunset, others just after. Some signal for only a half hour and others may signal for several hours. Each species also tends to restrict its activity to certain habitats, such as swamps, or woods, or open meadows.

Some other insects can produce light, but the fireflies are the only insects that flash it on and off in distinct signals. The larvae of fireflies also have the ability to glow, but why they do is not yet clear, since they have no mating needs at this time. Both the larvae and adults are carnivorous and eat other insects, as well as snails and other small ground creatures.

In late winter and early spring, you are likely to see some adult fireflies active during the day on the sunny sides of trees. This species, *Lucidota atra*, overwinters in the crevices of tree bark and moves about on warm days. It does not have a flash signal, for it would be useless during daylight.

TORTOISE BEETLES

Relationships

Tortoise beetles are a subfamily (Cassidinae) of insects in the family Chrysomelidae, or leaf-eating beetles, which in turn is in the order Coleoptera, or beetles. They are known as tortoise beetles because their wing covers and upper body

Tortoise beetle habitat with bindweed

Tortoise beetle adult on bindweed leaf. Life size

covers are oval and are flattened out at the sides, which makes these insects look like minute turtles or tortoises. Many of them are brilliantly colored with golds and iridescentlike greens, all of which fade when they die. Many species in both larval and adult forms feed on plants in the morning glory family.

Life Cycle

Tortoise beetles overwinter as adults in the leaf litter at the base of their food plants. They emerge in early to midsummer and begin to feed on leaves mostly from plants in the morning glory family. After mating, the females lay eggs either singly, or in clusters of ten to twenty each, on the undersides of leaves. The eggs hatch in one to two weeks and the larvae immediately start eating the leaves, while remaining underneath them. The larvae mature for four to six weeks and then pupate, usually while attached to the undersides of leaves. Pupation lasts about a week and then the adults emerge and feed for a few weeks at the end of summer. Then they leave the food plants and overwinter in dry protected places.

Highlights of the Life Cycle

The larval and adult stages of this insect are the highlights of its life; the adult stage because of its beauty and the larval stage because of its marvelous habit of carrying its feces and shed skin in a little packet held over its back. The egg and pupal stages can also be seen.

How to Find Tortoise Beetles

The easiest way to find these beetles is first to look for the plants they like to eat. The most common of these is bindweed, which is a herbaceous vine with heart-shaped leaves and large pink or white trumpetlike flowers. It is very similar to, and in fact related to, morning glory. It grows in moist, sunny areas, usually climbing up over other vegetation. Once you have found this plant, look for holes chewed in the leaves, and then near the holes look for oval, flattened beetles about one-quarter to one-half inch in diameter. They are often shiny colors of gold or red and glisten in the sun as they fly from areas where they are disturbed.

What You Can Observe

ADULT BEETLES

The adult beetles are almost round in shape and vary in size from one-quarter to one-half inch in diameter. Their most pronounced feature and distinguishing characteristic is that the hardened plates that cover the upper sides of their bodies flatten out at the edge of the beetle, like a little

Various species of adult tortoise beetles

shelf. This characteristic gives them a slight resemblance to tortoises. Usually their pronotum is so extended that it covers the head.

Some species of tortoise beetles are among the most beautifully colored of all North American beetles, especially the golden tortoise beetle, which looks like liquid gold and seems to change color in the sun. Some other tortoise beetles have spots and resemble ladybird beetles; a few others have stripes. The adults, like the larvae, chew many small holes in the leaves, leaving them with a riddled appearance. The adults fly very quickly from leaf to leaf, so may be hard to spot at first. They are likely to be seen either eating, mating, or laying eggs.

THE LARVAE

If you see holes in the leaves but can't find the adults, then you are in for a real treat, because this means that the beetles are in the larval stage, the most intriguing stage of their lives. To find them, check the leaves with holes for little dark flecks of what looks like just bits of junk. What you are seeing are collections of droppings and shed skins from the larvae. Underneath this "junk" is the larva itself, since it has the unusual habit of holding this collection of droppings and shed skins over the top of its body.

If you could see the larva without the junk, you would notice two long prongs extending from the tip of its hind end. When the larva is about to shed its skin, it attaches these prongs to a leaf. Its skin peels back from its head and finally ends up attached only to the prongs. The larva then detaches itself from the leaf and continues feeding, with the prongs holding the shed skin over its back. The larva also has an extended anal opening, which deposits feces onto the shed skin. This collection of stuff has been called

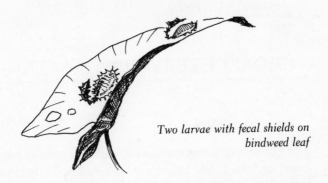

*Two larvae with fecal shields on
bindweed leaf*

a "fecal shield"; this name gives a slight hint as to its function.

Here is an experiment that you can perform to observe the uses of the shield. Look under the shield at the larva and you will see little branched spines all around the outer edge of its body. If you touch one of these lightly with a pine needle or grass blade, the larva will respond by swinging its fecal shield over to the point where you touched. It is believed that this is one of the purposes of the shield; when a predator, such as an ant, approaches the larva and starts to explore it, the larva presents the predator with the fecal shield. The observed effect has been that the predator generally loses interest and goes away, or if it comes in contact with fresh droppings from the larva, goes away while wiping its antennae clean.

Two other stages of these beetles can also be seen on the leaves. Turn the leaves over and look for the tiny, oblong, white eggs or small, dark pupal cases. After the adults emerge from the pupae, they hibernate in leaf litter beneath the plants and emerge the next spring to feed and lay eggs.

CHOKECHERRY TENTMAKER

Relationships

The chokecherry tentmaker is a species (*Archips cerasivoranus*) of insect that belongs to the family Tortricidae, or leaf rollers, which in turn is in the order Lepidoptera, or moths and butterflies. This is a family of small moths that are most conspicuous in their larval stage, for it is then that they have the habit of rolling leaves together with silk and feeding from inside them. Some, like this species, make large web nests. The adult, when at rest, closes its wings over its back, creating an outline similar to the shape of a bell.

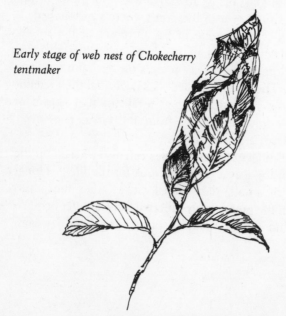

Early stage of web nest of Chokecherry tentmaker

Chokecherry tentmaker habitat

Life Cycle

Chokecherry tentmakers overwinter as eggs that are usually laid in a cluster at the base of their food plant, chokecherry. The eggs hatch in late spring, and the larvae crawl up to the leaves and tie them together with silk as they feed on them. In five to six weeks, the larvae crawl to the center of the web to pupate in loose-knit cocoons. Pupation lasts two weeks, at which time the pupae wriggle out of their cocoons and to the edge of the web nest, where they attach themselves. Within a day or two the adults emerge from the pupae. They fly about in late summer, mate, and then the females deposit clusters of eggs on the trunks at the base of chokecherry trees.

Highlights of the Life Cycle

The larval stage of these insects is the easiest to find, and through observing them over the course of their lives, you can see many of their unusual habits. Even finding the web nests that are empty late in the season can be fun, for you can see so much of the larval and pupal life history in them.

How to Find Chokecherry Tentmakers

In early to midsummer, the web tents of the larvae become conspicuous, and looking for these is the best way to find the insects. The web nest can be found on a great variety of fruit trees but is most common on chokecherry, a shrub relative of black cherry, that grows along roadsides and in waste areas. The web nests are at the tips of branches, and consist of leaves pulled close to the branch with silk.

What You Can Observe

DISTINGUISHING KINDS OF WEB NESTS

There are three different kinds of webbed caterpillar nests that are commonly seen from spring through fall. They are made by the tent caterpillar, the chokecherry tentmaker, and the fall webworm. Their nests can be easily distinguished by their forms and the seasons in which they are seen. The tent caterpillar web is found in spring in the crotch of forked branches; it does not contain leaves. The chokecherry tentmaker nest is seen mostly in summer, is made primarily on chokecherry, and contains leaves. The fall webworm nest is found in fall, is made on a variety of larger trees, including black cherry, ash, apple, and contains leaves. The tent caterpillar and fall webworm are described elsewhere in this guide, for they are most prominent in other seasons.

LARVAL HABITS

In early or midsummer, look for the nest of the chokecherry tentmaker. In late spring the eggs of this moth hatch from clusters that have been laid on the trunk at the base of chokecherry trees. Together the larvae crawl to the tip of one of the branches and pull the leaves close to the stem by attaching them with webbing. This tends to result in long, thin, and pointed webs. The caterpillars rest in the center of the nest during the day and feed on the leaves enclosed in the nest during the night. As the leaves are eaten, more leaves and branches are tied into the nest. Since chokecherries are generally small shrubs, the whole plant can be defoliated, in which case the caterpillars begin to extend their nest over grasses and other plants as they search for more chokecherry trees.

Once the larvae have fed for five to six weeks and are

Nest with pupal cases hanging on outside

Adult moth

full grown, they move to the center of the web and make loose-knit cocoons in which they pupate. After two weeks, the pupae wriggle out of the cocoons and move to the outer edge of the web nest where they attach themselves and hang down. This is one of the interesting variations between the life of this insect and that of either the tent caterpillar or the fall webworm, both of which leave the webbing and form cocoons in other protected places. So finding evidence of cocoons inside, or pupae on the outside, of a web nest is a good clue to the presence of the chokecherry tentmaker.

THE ADULTS AND EGGS

The adult moths emerge from the pupae in June and July. They are small, inconspicuous, brown moths with a wingspan of a little less than an inch. Like most moths, they are nocturnal. They mate and lay their eggs at the base of their food plants. The egg masses are hard to see. They contain thirty to one hundred eggs but are covered with a gray-brown, shiny substance that makes them blend in with the tree bark. The best way to find them is to locate a plant with the web nest and then look for the egg cluster (with already hatched eggs) on the lower portions of the plant. The insect overwinters in the egg stage and hatches in spring when the new leaves are emerging on the chokecherries.

MIDGES

Relationships

Midges are a family (Cecidomyidae) of insects in the order Diptera, or flies. They look almost exactly like mosquitoes at first glance, but with a second look you can easily distinguish them. And there is a good reason to, since they are harmless and don't bite.

The larvae are for the most part aquatic but a few live under bark, or in soil or rotting material. The adults are often near water and form large swarms that have fascinating behavior.

Life Cycle

Midges overwinter as larvae underwater in mud or decaying vegetation. In spring the larvae continue to feed on minute

Midge habitat

plants and animals and then rise to the water surface as pupae. The adults crawl from the pupae onto the surface of the water and fly off. Adults form large swarms in which mating takes place, after which fertilized females spend one or two days laying eggs. The eggs are placed in the water in long connected strands. The eggs hatch in a day or two, and the larvae, for the most part, construct tubes of debris around them in which they mature and feed. They mature to adults in two to four weeks. There are many broods and the larvae of the last brood overwinter.

Highlights of the Life Cycle

The larvae are common, easily found, but are small and hard to observe. The adults live only a day or two and do

Midge swarm. Life size

no feeding, but during that time they have some incredible behavior with respect to swarms, which are a behavioral adaptation for mating and finding mates. Studying swarming behavior, especially with these nonbiting midges, can be extremely entertaining.

How to Find Midges

Adult midges are easiest to find when they are swarming. Swarms are most common in the morning and evening but occur in midday as well. They can be found anywhere, but especially good places are at the edges of water. As you become aware of midge swarms, you will also begin to see swarms of other types of flies, and these are often formed just over the top of a prominent object, such as a tree in a meadow or even your own head.

What You Can Observe

SWARMING BEHAVIOR

Most people, when they see a swarm of flies, assume two things: that they are all mosquitoes and that they are about to bite you. Neither is true. Thousands of species of flies form swarms and only a very small proportion of these bite humans at all. And even if the swarm is of mosquitoes, chances are that you will not be bitten, since most swarms are composed only of males and the males do not bite.

The majority of swarms you see are likely to be composed of nonbiting midges, so there is no danger in getting near them to watch. If you are like me, you probably never wondered why some flies swarm, but just figured that it was something flies did and left it at that. But if you take the time to watch, you will find that there is much more order and predictability in the behavior of these swarms than you ever expected, and that seeing the patterns of behavior in swarms can be great fun. The swarming behavior of midges is described here, but in fact most of their behavior can be applied to other types of fly swarms that you will see as well.

Midge swarms tend to be localized both in place and time. If you see one at a certain time and in a certain place, chances are good that you will see it the next day at the same time and same place. Most midges have specific times of the day when they swarm. It may be twice a day, in the morning and evening, or it may be just once, maybe in the middle of the day. The timing for each species is believed to be either triggered by a certain level of light, or regulated by an internal clock. Each species has its own fixed pattern of timing.

Why swarms form where they do is not as easy to understand. You might have already seen swarms over water, in patches of afternoon light, or even right over your head. It has been found that certain species of flies respond to specific aspects of the environment, called "swarm markers," and always form swarms in these or very similar circumstances. In some cases, researchers have even been able to mimic enough aspects of the swarm marker to actually cause flies to swarm over certain spots. For instance, a common midge will swarm over shiny black plastic at least four inches by four inches, placed on the ground. Some other flies swarm just above the top point of a tree,, and others just beneath the tip of a branch. I find that many swarms in forests are often located right in the center of a patch of light, making them especially conspicuous.

A swarm also typically stays within a certain distance of its marker. This is believed to be the result of the limited vision of the insects. At some point from the marker they can no longer see or discern its significant factors. This is the maximum distance they can move away from it without losing it. Thus, in the absence of wind, the swarm will generally stay at about the same height fairly constantly. But you will also see the swarm drop down and even out of sight and then reappear again. This is almost always in response to breezes. As the wind gets stronger, the swarm moves lower so as to be able to remain over or under the marker and not get blown away. As the breeze subsides, the swarm rises. On very windy days, flies may be unable to swarm.

If there is a slight breeze, all of the flies in the swarm face the wind. They often seem to fly forward, possibly to the windward limit of the marker, and then drift back with the wind, possibly to the leeward limit of the marker, and then fly forward again. This creates a kind of circular motion, which has often led to people describing the flies as "dancing." There are other types of movement that you will see among the flies in a swarm, many of them seemingly synchronized, and their function is still unknown.

Midge antennae. Left: *male*; right: *female*

One problem for midges in a swarm is being able to distinguish females from males. Studies have shown that the antennae of midges play a key role in sexual recognition. This is also true of mosquitoes. The antennae of the males in these species are long and extremely plumose, and at their bases have special organs, which, along with the antennae, pick up vibrations in the air. The females give off a buzz of a certain pitch, or range of pitches, when they are sexually mature, and the males are attracted to this pitch when it is within about a foot of them. Their own flight buzz is lower and out of their own "hearing" range. The flight buzz of certain common female mosquitoes is around middle C on the piano. Singing this pitch near a swarm of male mosquitoes will make them all immediately fly in your direction.

The main reason for this is at the heart of the chief function of fly swarms. In each species, male and female flies are internally programmed to swarm over a given marker at a fixed time of day. As the females enter the swarm, males attach to them and the pairs leave the swarm and mate, usually nearby. Most swarms are composed almost entirely of males, since as soon as a female enters the swarm she tends to join with a male and leave to mate. Once she is mated she will then go off and start egg-laying, whereas the males are more likely to return to the swarm. Thus, swarms seem to help males and females of a species locate each other for mating.

There are still many unanswered questions about swarming behavior. Maybe some of your own observations of fly swarms will add needed information that will help solve this fascinating puzzle.

ROBBER FLIES

Relationships

Robber flies are a family (Asilidae) of insects in the order Diptera, or flies. The adults all prey on other insects, usually catching them in midair and then sucking out their juices. Many look much like bees and wasps; you must look closely to recognize them as flies. They have long legs with which they capture and hold prey. They are extremely common and often beneficial in controlling insect pests.

Life Cycle

Robber flies overwinter as larvae under the soil. The larvae are like tiny, slightly flattened worms. In spring they resume crawling about in the soil, feeding on decaying vegetable matter and on the larvae of other insects. In early summer they pupate in the soil and, when just about to end pupation, wriggle up to the surface so that half the pupa is

Bee-killer robber fly feeding on captured bee. Life size

exposed. The adult then emerges from the pupa. Adults feed on other insects, mate, and the fertilized females lay eggs on or just beneath the soil surface. The larvae immediately start to crawl about in the soil and feed. In some species the larvae remain beneath the soil, feeding for an additional year. In either case, it is the larval stage that overwinters.

Highlights of the Life Cycle

The adult phase is the only one that is visible, but this is no loss, since it provides more than enough entertainment for even the avid insect watcher. Watching the adults fly out from their perches and try to catch other insects is great fun. They also have unusual mating displays between the sexes, involving flights and different pitches of buzzing. For most species the mating behavior has yet to be studied.

Robber fly habitat

How to Find Robber Flies

At the borders of woods, fields, or water, look for long-legged, bristly flies perched on leaves or twigs. If you look closely you can see their heads move to follow the flight of other insects. Every so often the fly will dart out after a passing insect and then return to its perch, sometimes having caught its prey. Many robber flies look like bees or wasps, but no bees or wasps have this type of behavior.

What You Can Observe

HUNTING BEHAVIOR

Hunting is done from perches that are usually located on sunlit leaves, twig tips, or, in some species, even occasionally on the ground. The fly perches, looks for insects flying by, and follows their movements by tilting its head and quickly reorienting its body. The robber fly has a very mobile head, and it moves around a great deal as the fly follows the movements of other insects. In fact, if you flick a small pebble or bit of twig over a robber fly, it will follow it visually by moving its head in the direction of the object.

The foraging range of most robber flies from a perch is about five yards, but the majority of insects are caught within one yard.

There are basically two types of robber fly flights: contact flights and no-contact flights. In the former, the fly actually tries to catch an insect; in the latter, a robber fly investigates a prey insect, decides that it is not right, and returns. Robber flies do not seem to be able to distinguish suitability of prey except at a short distance, and so will fly out to investigate even bits of falling leaves, twigs, seeds, et cetera.

It is interesting to think about how the robber fly intercepts the flight path of its prey. This is quite an ability, for

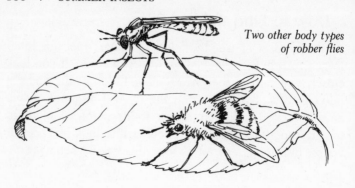

*Two other body types
of robber flies*

the fly rarely chases its prey, but rather intercepts it in midair, so the fly must judge the speed, trajectory, and distance of its prey. If you have ever tried to throw a ball to a person who is running, you are familiar with how hard this can be. Some robber flies have been observed to compensate for discrepancies in the prey's speed by arriving at a spot in the path of the prey, hovering there for an instant until the prey passes, and then capturing it. The prey is caught in the legs of the robber fly in much the same manner that dragonflies catch prey, and it is paralyzed when the fly, still in the air, sticks its mouthpart into the back of the prey and injects a paralyzing chemical. The robber fly then goes back to its perch with its prey impaled on its mouthpart.

When the fly lands on its perch, it may move back into the shade, and then it starts eating by sucking out the insides of its prey. This eating may take anywhere from five to thirty minutes. During this time the fly may need to readjust the position of its prey in order to eat it more effectively. To do this, it flies up with the prey and while hovering adjusts the prey with its legs. Then it lands and continues feeding. When the fly is finished feeding, it drops the remains of its prey on the ground.

If the fly does not catch prey from a particular perch in five to ten minutes, it may go to a new perch, usually about ten to fifteen yards away. Only about 15 percent of all flights from a perch result in the capture of prey. The main prey of most robber flies include other flies, wasps and bees, and beetles.

COURTSHIP

While watching robber flies hunt, you also may see some behavior associated with courtship. Courtship of predatory insects has its risks for both the male and female, since there is always the possibility of being treated as prey rather than partner. Possibly in response to this, the males, when approaching females, do a number of visual displays that may trigger new responses in the female and curb her tendency to treat the male as prey. The visual displays of male robber flies include hovering in front of the female, sometimes accompanied by high-pitched buzzing; or various wing and/or leg movements in front of the female. The female usually responds with wing and/or body movements of her own.

SYRPHID FLIES

Relationships

Syrphid flies are a family (Syrphidae) of insects in the order
Diptera, or flies. The adults frequent flowers and feed on
nectar. Next to bees, they are our most important pollinators of flowers. Many of the species are very similar in
appearance to bees or wasps, and since they are usually
near flowers, their behavior is similar as well. One difference in behavior is that syrphid flies are excellent hoverers
and are able to remain absolutely stationary in the air.
When bees and wasps hover, the best they can do is to bob
up and down in one area.

Life Cycle

Syrphid flies overwinter as larvae. In spring or early summer
the larvae pupate. Pupation lasts one to two weeks in most
species. The adults emerge in summer and feed on flowers.

Syrphid fly on daisy. Life size

Syrphid fly habitat

Depending on the species of syrphid, the female lays her eggs near aphids, near plant bulbs, or on decaying flesh or dung. Each of these habitats requires special adaptations in the form and behavior of the larvae that live in them. The larvae feed in the summer and usually go into the soil to overwinter.

Highlights of the Life Cycle

Syrphid flies are particularly interesting to watch as they hover about flowers, feeding, and in some cases defending territories. They are easy to find on flowers, and, like so many of our common and beautiful insects, most species have not been studied for their behavior; in fact, for many, even their life cycles are not known.

How to Find Syrphid Flies

Adult syrphid flies can be found wherever there are flowers. They are usually colored with black, yellow, and orange markings on their abdomens and, because of this, look just like some species of bees. They can be distinguished from bees by these clues: they have two wings, bees have four; when at rest their wings are held slightly out to the sides, bees fold them over their backs; their antennae are shorter than the length of their heads, bees' antennae are considerably longer than the length of their heads.

What You Can Observe

ADULT BEHAVIOR

An enjoyable activity on bright summer days is going to a garden, a patch of wildflowers, or a shrub in full bloom, and looking for syrphid flies among the bees and wasps. There are almost sure to be a number of both groups of insects. You may never have heard of syrphid flies, but chances are you have seen them, for they are one of our most common flower visitors.

The adult syrphids come to flowers to feed. They live

Various species of adult syrphid flies

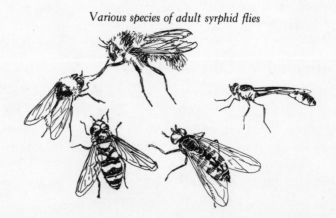

almost entirely on nectar, but occasionally lap honeydew from aphids off the surface of leaves. As they collect nectar, they inadvertently get pollen on their legs and bodies and carry this on to the next flower. In doing this they help pollinate the flowers. Something you may notice when a syrphid fly lands on a flower is that its wings stop moving but its buzz keeps buzzing. This curious event is believed to be the result of continued contractions of its thorax, while the wings are, in a sense, "out of gear."

As mentioned earlier, syrphid flies are expert hoverers, able to remain absolutely stationary in midair. This ability is even more remarkable when you think about all the little gusts of wind that are swirling about the fly as its body remains motionless. It is believed that it does this through sight, fixing its position in relation to an object in its environment.

MATING AND TERRITORY

It is known that some flies, especially the males of certain species, hover in certain spots to attract the attention of females. It has also been discovered that some species of syrphid males patrol groups of flowers, especially in the morning when females are coming to them to feed. They continue around on a route as long in circumference as one hundred yards and stop to hover in front of the flowers, rest on the leaves, or feed at the flowers. When they run into females on their route, they fly after them and attempt to mate. They also chase away other insects that they encounter on the route. Their route may overlap with that of another male, but as long as they do not meet, this does not matter. They have been observed to complete their route three to five times an hour and to be seen following the same route for up to three days.

In the afternoon, the males of some of these species have

been found hovering about areas where females lay their eggs, such as in the collected water of tree holes. They remain in an area about two yards in diameter. When fertilized females arrive, they mate with them. This is effective because in species of flies where the female mates more than once before laying eggs, the sperm from the last mating fertilizes the majority of eggs.

LARVAL HABITS

While almost all of the syrphid adults feed on flowers, their larval habits are more varied. The larvae of most of the colorful syrphids that we see in our gardens are aphid-feeding larvae. The adult lays its eggs on branches or leaves where aphids feed, and the larvae prey on the aphids. This is described more fully in the section on aphids and their predators. Many other syrphid larvae live in the water or in juices of decaying animal or vegetable matter.

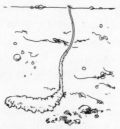

Larva of drone fly (rat-tailed maggot) buried in putrid water

One of our common species in the garden, often confused with bees, is called the drone fly, and its larva has some amazing habits in regard to living in water. It often grows as a larva in puddles around the base of manure piles. In order to breathe while feeding underwater, it has a tube at its rear end which sticks just out of the surface. This has a little ring of hairs that splay out over the surface of the water to keep the tip afloat. What is even more remarkable is that this tube can extend itself, like a telescope, up to four inches and then retract to accommodate different depths of water that the larva feeds in.

RELATION TO BEES AND WASPS

Why do syrphids look so much like bees and wasps? In some cases the answer seems clear. There is a group of these flies in the genus *Volucella* whose females crawl into the nests of bees and wasps and lay their eggs. The larvae then feed on the remains of the dead bees and wasps that fall to the bottom of their dwellings. In these cases, the adult flies may look very similar to the species of bees or wasps for camouflage. But what about the other hundreds of species that never get near the homes of wasps or bees? How has it been to their advantage to get this coloration? The answers to these questions will be gotten only through careful observation of the flies' behavior, and, amazingly, even though these flies are some of our loveliest and most common backyard visitors, only a few species have been studied.

Fall Insects

Observing Insects in Fall

The best place to observe insects in fall is on the flowers of goldenrod. Many of our most common insects are attracted to the large amounts of pollen and nectar that these offer. A visit on a warm sunny day will enable you to watch honeybees, longhorned beetles, soldier beetles, and paper wasps, all feeding. Just as common, but well camouflaged, is the marvelous ambush bug, which sits and waits to catch and feed on one of the other insect visitors. Also on goldenrod blossoms will be insects mentioned in other seasonal sections, such as the white and sulphur butterflies and bumblebees (see the spring section), the leafhoppers and syrphid flies (see the summer section), and the mantids and viceroy butterfly (see the winter section).

While still among the goldenrod, keep your eyes open for a monarch butterfly gliding about just above the plant tops. Also take a moment to listen to the sounds of the crickets and grasshoppers that are bound to be making their calls all about you. See how many different types of calls you can distinguish.

As you leave the goldenrod patch and get back on a path or road, stay on the alert for woolly bear caterpillars, which you are sure to find crossing roads or paths in search of a proper overwintering site. And if you go beneath an oak, pick up some acorns and look for holes in them or test them for softness, either of which may lead you to discover the work of acorn weevils, also very common insects. At the tips of tree branches, you may see masses of webbing enclosing the leaves. These are the nests of fall webworms. If you pass pines, look along the tips of their branches for groups of caterpillarlike larvae feeding on the needles. These are most likely the larvae of sawflies, but they are less commonly found than other insects mentioned above.

Finally, if you are at the edge of a pond or other area of water, look for the white fluff of the woolly aphids along the stems of alder; and in areas of still water, see if you can spot backswimmers suspended at a forty-five-degree angle just beneath the water surface.

These are the insects best observed in fall. The easiest to find were mentioned first; those less conspicuous were mentioned last.

FALL INSECT LOCATION GUIDE

IF YOU ARE NEAR LOOK FOR	FIELDS	PONDS & STREAMS	WOODS	FIELD EDGES	HOUSES	BARE GROUND
CRICKETS AND GRASSHOPPERS	■		■	■	■	■
BACKSWIMMERS		■				
AMBUSH BUGS	■			■		
WOOLLY ALDER APHIDS		■		■		
LONGHORNED AND SOLDIER BEETLES	■					
ACORN WEEVILS			■			
MONARCH BUTTERFLY	■					
WOOLLY BEAR						■
FALL WEBWORM			■	■		
HOUSEFLIES					■	
SAWFLIES			■	■		
HORNETS AND YELLOW JACKETS	■				■	
PAPER WASPS	■				■	
HONEYBEES	■			■		

CRICKETS AND GRASSHOPPERS

Relationships

Crickets are a family (Gryllidae) of insects in the order Orthòptera. *Grasshoppers* is a general term applied to several families also in the order Orthoptera. All members of this order have two pairs of wings: the first pair are thin, leathery strips that cover the last pair, which open and close like the pleats of a fan. Most of the insect sounds we hear in late summer and fall are made by crickets and grasshoppers. The sounds are usually made by the males and help attract females.

Life Cycle

The life cycles of crickets and grasshoppers are similar. Most overwinter in the egg stage. Nymphs hatch from the

Short-horned grasshopper above and field cricket below. Life size

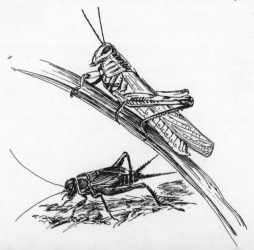

Cricket and grasshopper habitat

eggs anywhere from late spring to late summer. Grasshopper nymphs mature in about sixty days and undergo about five molts. Cricket nymphs take sixty to ninety days to mature and undergo eight to twelve molts. Adult crickets and grasshoppers make sounds to attract mates. Some even defend territories. Fertilized females generally lay eggs in late summer and fall. Many grasshoppers lay small packets of ten to twenty eggs in the soil. Crickets tend to lay their eggs singly, either in the soil or inserted into plant stems. Most grasshoppers and crickets have only one brood per year, but in warmer climates some, like the field cricket, may have up to three broods and may overwinter as nymphs and adults as well as eggs.

Highlights of the Life Cycle

Certainly, for us, the highlight of these insects' lives is their adult phase, when they fill the late summer and fall air with their musical sounds. Each species has a distinct call, which is used to attract mates and keep away competitors. Grasshoppers and crickets have very different styles of making sounds and of mating behavior, and these are fun to know about and fairly easy to experience.

How to Find Grasshoppers and Crickets

The easiest way to find grasshoppers or crickets is to listen for all chirps, trills, or mechanical sounds that are rhythmically repeated. Listen in areas of dense vegetation, for the insects feed on plant matter. Also, listen at different times of day and night, since some species call only at very specific times.

What You Can Observe

LISTENING TO THE CALLS

The most prominent feature of insect behavior in late summer and fall is undoubtedly the constant calling of crickets and grasshoppers. It is both easy to hear and easy

Male field cricket calling

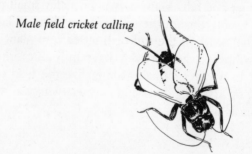

to forget. For most of us, it soon ceases to be part of conscious hearing and is relegated to the status of a background noise. This fall, try to reverse the process; bring this wealth of sounds back into your consciousness and take some time to enjoy its diversity and behavioral significance.

The first broad distinction to be made is between sounds that have a musical pitch (try to hum a matching note) and songs that are purely mechanical and have no pitch to hum along with. The former often sound like whistles or trills, the latter sound like pieces of sandpaper being rubbed together. This distinction will help you sort out which animals are making which sounds. Sounds with pitch are made by the crickets. Sounds that are pitchless and mechanical are made by the short-horned grasshoppers and long-horned grasshoppers.

Each species of cricket and grasshopper has a different song, and you can come to recognize individual species from their songs, just as you would birds. An enjoyable activity is to take a walk on a warm, sunny day and see how many different types of songs you can hear. Each species tends to live in a particular environment and give its calls at specific times of the day or night. Therefore, to hear the greatest variety of calls, visit different habitats at varying times of day. Good places to hear these insects are along roadsides and in woods or fields. Here are some ways in which calls vary. They may be short calls, continuous trills, or a mixture. Their rhythm may be steady or irregular. And they may be loud or soft, high- or low-pitched. Each call is generally from a different species of grasshopper or cricket.

THE FUNCTIONS OF CALLS

Most grasshopper and cricket calls are made by the males. They give a call to attract the females of their own species.

Field crickets. Left: *female*; right: *male*

This call is often referred to as the "song," and it is generally the longest and most complex sound that the insect produces. The female responds to a male's song by moving toward the source of the sound. In some species, the female has a song as well, and this is alternated with the song of the male; here male and female move toward each other. Once males and females are within a few inches of each other, softer sounds may be produced by the male, which have the effect of stimulating mating behavior in the female. These are often called "courtship calls." These quieter calls are less often heard. Once the two have mated, both usually cease all calling until they are ready to mate again.

Cricket and grasshopper calls serve two other functions: "aggressive calls" are used between males and generally make the callers move farther apart; and "alarm calls" often cause all other insects in the area to stop calling for a brief period, possibly for protection from predators.

WATCHING THE INSECTS

You don't have to rely just on hearing crickets and grasshoppers, for they are easily seen and you can also watch much of their behavior. The three main types we have mentioned — crickets, long-horned grasshoppers, and short-horned grasshoppers — are easily distinguished and include most of the grasshoppers and crickets you will find. The short-horned grasshoppers, sometimes called locusts, form one group and are recognized by their short antennae (less than half the body length). The long-horned grasshoppers and crickets form another group and are recognized by their long antennae (body length or longer). Females of the latter group have long, swordlike ovipositors, while female short-horned grasshoppers have blunt-tipped abdomens. Long-horned grasshoppers and crickets can be distinguished by the arrangement of their wings. Long-horned

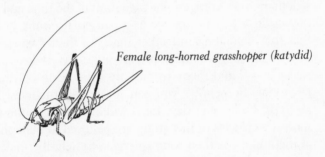

Female long-horned grasshopper (katydid)

grasshoppers hold their wings in a tentlike fashion over their backs, whereas crickets fold their wings flat over their back.

The two groups produce sound in different ways. Short-horned grasshoppers rub a series of small projections on their hind legs across a scraper on their wings, much like running a comb across your fingernail. The long-horned grasshoppers and crickets rub a series of ridges on one wing against a scraper on the other wing. Both of these types of sound production are very different from the way cicadas produce sound (see Cicadas, summer section).

MATING BEHAVIOR

The behavior of these two groups is also quite different. Male short-horned grasshoppers are active and mobile and frequently do short vertical flights, during which they make a crackling sound. These flights are common on hot days along sunny dirt roads or paths. The songs of the males tend to make other males come to the same spot, where they then compete for a female that is also attracted. After the males get close, you may see them lift a hind leg up and down, making no noise. This may be a communicative display that helps them to determine dominance and decide which will mate with the female.

Male long-horned grasshoppers and crickets are more sedentary, remaining in one area most of the time and even staying in a tiny burrow or in a particular clump of grass that they defend against other males of their own species. The male song in these insects may in fact keep other males away, and many species are believed to be strongly territorial. Sometimes you can locate a particular cricket and visit it day after day, for it will always be in the same spot. The insects of this group are believed to mate mostly at night and use their long antennae to help the male and

Flying short-horned grasshopper (Carolina locust)

female orient themselves to each other. On the other hand, the short-horned grasshoppers mate mostly during the day and orient themselves mostly through visual cues. This

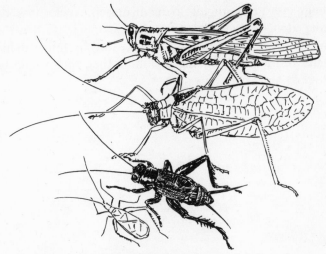

Top to bottom: *Short-horned grasshopper, long-horned grasshopper, field cricket, tree cricket*

may be part of the explanation for why each group has its own antennal length.

CHORUSING BEHAVIOR

Many of these insects have chorusing behavior, in which they all sing at once in a given area. In some cases they synchronize the rhythms of their calls. This is especially true of the snowy tree crickets, which can be heard in fall giving synchronized choruses of short, low, musical chirps in even beats from among grasses and shrubbery. As with all of these insects, the calls of the snowy tree cricket tend to slow in colder weather and speed up in warmer weather. The cricket is sometimes called the temperature cricket, for by counting the number of its chirps in fifteen seconds and adding it to 37, you get an approximation of the temperature in degrees Fahrenheit.

There are also instances of male crickets alternating their

songs instead of singing them in unison. When they do this, though, each male sings half as much, so that the overall effect is the same as one male singing. If they didn't make this adjustment, the rhythm of their song would be twice as fast as normal.

BACKSWIMMERS

Relationships

Backswimmers are a family (Notonectidae) of insects in the order Hemiptera, or half-wing bugs. Most species are about a half inch long and spend almost all of their life underwater. Their common name refers to their habit of swimming upside down, a unique characteristic. They have a sharp, beaklike mouth, which they poke into other animals they have caught and through which they suck out the prey's juices.

Life Cycle

Backswimmers overwinter as adults underwater and are sometimes active right through winter. In spring the adults mate, and the fertilized females lay slender white eggs and attach them to underwater plant stems. The eggs hatch in two to three weeks and the nymphs develop into adults in about two months. Adults can fly and may move to new ponds or areas of water before laying more eggs. There are one or two generations per year, and the adults of the last generation overwinter.

Backswimmer adult suspended from water surface. Life size

Highlights of the Life Cycle

In their adult stage, when they are largest, backswimmers are most easily observed. While the insects are still in their natural habitat, you can enjoy seeing several features of their behavior, such as locomotion, breathing habits, and prey-catching. Features of their anatomy can be better appreciated by putting the insects temporarily in a water-filled glass jar, and afterward returning them to their home. Backswimmers can inflict a painful bite, so avoid handling them with your bare hands.

How to Find Backswimmers

Go to the edges of still ponds, lakes, or the backwaters of slow-moving streams and look for dark insects about one-half inch long, suspended at a forty-five-degree angle just beneath the water surface. This position is unique to and typical of backswimmers. You may also see the insects

Backswimmer habitat

swimming in a jerky manner, their hind pair of legs clearly much longer than all the rest.

What You Can Observe

BACKSWIMMERS AT THE WATER SURFACE

In just about any clear pond or pool you will find back-swimmers: small, oval bugs about a half inch long, with one pair of legs extended out and angling toward their heads. You can recognize them by both their resting positions and movements. They have a habit of resting at the water surface with the tips of the abdomens just breaking the surface and their heads hanging lower at about a forty-five-degree angle. The long legs are used like oars and propel the insect forward in sharp jerks, rather than in a continuous movement.

Backswimmers hanging at the water surface may be doing either of two things. They may be gathering oxygen. They have three longitudinal rows of hairs on their bellies, forming two compartments that trap air in them for when the insect dives underwater. These are periodically emptied and refilled at the surface. Backswimmer nymphs must replenish their oxygen every three to five minutes, but the adults can remain underwater without refilling for hours.

Backswimmers at the water surface may also be searching for prey. They eat other insects, tadpoles, small fishes, and small crustaceans, and are believed to detect through hairs on their legs vibrations made in the water by their prey. They then orient themselves toward their prey, dart after it, and catch it with their front legs. Immediately, the backswimmer inserts its beaklike mouth into the prey and pumps in digestive juices that kill the prey and break down its inner structure. The backswimmer then sucks out the

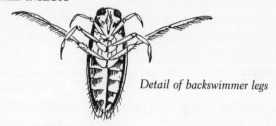

Detail of backswimmer legs

digested innards. The large eyes of the backswimmer are believed to be used only for close-up orientation to its prey.

ADULT BEHAVIOR: SWIMMING, DISPERSING, MATING

Backswimmers are wary and often dart away when you approach. But since they carry so much air in the hairs on their abdomens, they automatically float to the surface unless they continually swim downward or hold on to stable objects in the water. These insects should not be picked up in your hand, since they can inject their digestive fluids into humans as well as into other insects, and this can be painful.

They are called backswimmers because they are actually swimming upside down all of the time. This can be clearly seen if you catch one and place it in a little bit of water. As you ladle the water out, it will finally get so shallow that the backswimmer will flip rightside up. As you add water again, it will flip over on its back. The wing covers on its back come together into a ridge that may act like the keel of a boat as the insect swims. Also notice that the protective coloration of the insect is reversed, for it is light on its back and dark on its belly, the opposite of most animals.

The males are known to make noise when they approach the females for mating. They do this by stridulating — rubbing the front legs against the base of the mouthpart.

Eggs are laid attached to plant tissues and even occasionally to other adult insects. The young nymphs are similar to the adults in appearance and behavior, but they cannot mate or fly. The adults disperse to new areas by flying in swarms at night, and they may be seen around lights at night. The common species overwinter as adults and remain active. If you find a hole in the ice of a still pond, look down and you may see them still swimming about in the middle of winter.

AMBUSH BUGS

Relationships

Ambush bugs are a family (Phymatidae) of insects in the order Hemiptera, or half-wing bugs. They are about a half inch long and are shaped like an hourglass. They often wait camouflaged among flowers for other insects to come near. Once the insect is near enough, the bug quickly reaches out with specially adapted front legs and catches its prey.

Ambush bug habitat

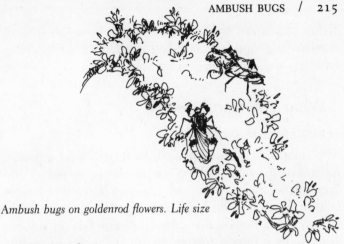

Ambush bugs on goldenrod flowers. Life size

Life Cycle

Ambush bugs overwinter as eggs placed on the stems and leaves of plants. In spring the eggs hatch and the tiny nymphs, no bigger than half the size of a fruit fly, catch and feed on insects twice their size. In one to two months the nymphs undergo five molts and are finally mature in late summer. They mate in late summer and fall, and the females lay the eggs, which overwinter. The adults die as cold weather sets in.

Highlights of the Life Cycle

Ambush bugs can be most easily found and observed in fall when they are adults and at their largest. Their predatory habits, described later, are among their most fascinating feature to enjoy.

How to Find Ambush Bugs

Look carefully among the blossoms of goldenrod for a well-camouflaged insect, about one-half inch long, colored with yellow and brown or black. It is roughly hourglass-shaped, slow-moving, and has clawlike parts on its front legs. It is harmless and can be picked up.

What You Can Observe

PREDATORY BEHAVIOR

The ambush bug is well named, for it waits among flowers for insects to come and feed on the nectar or pollen and then catches them. It does not restrict itself to yellow flowers (within which it seems best camouflaged) and can be found on any other flowers that attract a steady flow of suitable visitors. Prey insects include many of the flower-visiting flies, such as syrphids and tachinids; smaller bees, such as halictids and the honeybees; and butterflies.

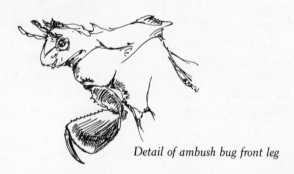

Detail of ambush bug front leg

It is safe to pick up one of these bugs and hold it with the tips of your fingers to get a closer look. One interesting feature is its front legs, which are adapted for capturing prey. At their tips is a fine, sharp blade that can snap back into a larger portion in the next section of the leg. The blade actually fits into a small groove. Although very small, this pincer is able to capture and hold insects much larger than the whole bug, such as white butterflies.

The ambush bug waits, nestled in a group of flowers. When an insect comes near enough, it grabs it on any available part: antennae, foot, or wing. It then inserts its beaklike mouth into the prey, usually at a weak point in the prey's body, such as where two portions join, and injects a fluid into the prey's body that both kills the victim and breaks down the structure of the interior organs and muscles. The pumping in of this fluid usually extends the body of the prey. Then the ambush bug sucks out the partially digested innards. After the ambush bug has eaten, just the external skeleton of the victim exists, and this is discarded. Sometimes these cast skins can be found among the flowers where ambush bugs have been feeding. In fact, it is not uncommon to spot a honeybee or fly that seems to remain curiously still, and then discover that it has just been discarded or is in the process of being eaten by an ambush bug.

MATING BEHAVIOR

Another common sight is two ambush bugs walking about, one on top of the other. This is not mating, but an activity that seems to precede mating. The male is on the back of a female. The two sexes can be easily distinguished, since the female is larger and bright yellow, while the male is noticeably smaller, slimmer, and darker. Sometimes the bugs may be feeding in this position, but if so, it is most often the female that both catches and feeds on the prey. The mating position is different, with the two insects side by side and the tips of their abdomens connected.

A COMMON NEIGHBOR

Often on the same plants as the ambush bug is a particular kind of spider that has many of the same predatory habits. It is called the crab spider because its front two pairs of legs are longer than the others and held out in front of the

Crab spider on goldenrod flowers

spider, giving it a slight resemblance to a crab. Also, it walks better sideways than forward and backward. The spiders are either light yellow or white, depending on the color of the flower they are on. They can change from one color to the other in about ten days. In either color phase, they have a red stripe on each side of the abdomen. In fall, crab spiders are found on goldenrod along with ambush bugs. They also wait in hiding among flowers for insects to come near and then reach out and capture them.

WOOLLY ALDER APHID

Relationships

Woolly alder aphids are a species (*Prociphilus tessellatus*) of insect in the family Eriosomatidae, or woolly aphids, which in turn is in the order Homoptera, or same-wing bugs. Many members of the family have the ability to produce from the backs of their abdomens strands of waxy material, which have the appearance of wool or cotton candy when the insects are grouped together. Like most other families of aphids, both the adults and nymphs feed on sap. Woolly aphids feed primarily on woody plants.

Woolly alder aphid habitat

Woolly masses on alder twigs.
Life size

Life Cycle

There are two types of life cycles in this species. In one, the aphids remain on alder trees throughout their lives. They are believed to overwinter as adults in the leaf litter at the base of an alder. In spring they crawl up the plant and feed on its sap. There are several generations per year and adults of the last generation overwinter.

In the other life cycle, the aphids alternate between two plants. The aphids overwinter as eggs placed on maple twigs. In spring they hatch into females, which feed on the undersides of maple leaves and reproduce. They are wingless, but in midsummer produce winged offspring, also females, which fly to alders. These females feed and reproduce on alders, and give birth to wingless young. Then in late fall, they produce winged young, which fly back to maples and give birth to both male and female young. The males and females mate, and each fertilized female lays a single egg on a maple twig. Only the eggs overwinter.

Highlights of the Life Cycle

The easiest time to see woolly alder aphids is in fall when the leaves are off the trees and the white wool of the aphids can be easily spotted on the dark alder trunks. A close look at a colony of these aphids shows you their wax-producing tubes and their winged and wingless forms. You may also find the caterpillar of the harvester butterfly feeding on the aphids, or other insects feeding on the excess honeydew produced by the aphids.

How to Find Woolly Alder Aphids

There are two places to watch for woolly alder aphids in fall: on the stems of alders and in the air. On the stems of alders they are very conspicuous, for they are bright white and look like a growth of sticky, cottony material surrounding portions of the stems for several inches. Alders can be found at the edges of wet areas. The airborne aphids appear on warm, sunny days in areas where there are either alders or maples, and they will look like tiny tufts of cotton floating up through the air.

What You Can Observe

APHIDS, "WOOL," AND HONEYDEW

If you find the aphids clustered on alder stems, take a moment to poke through the cottony material with your

Close-up of aphids and their "wool"

finger, and you will see the insects beneath. They are about an eighth inch long and look like miniature hand grenades, being oval with little projections all over their backs. Each one of these projections secretes a strand of the "wool," and together they conceal the insect and increase its apparent size to about a quarter of an inch.

Like all common aphids, these woolly aphids suck sap from the host plant. Part of this sap is used to create the waxy material for the wool, and the rest helps the insect to grow. Excess sap is excreted as honeydew, which sometimes falls on the leaves and branches below and makes them sticky. Two types of fungi often grow on branches below the aphids, and they are believed to feed off the honeydew. One type encrusts the branches with black material, and the other creates a large, spongy mass just below the aphid cluster.

OTHER INSECTS WITH THE APHIDS

Two other insects interact with woolly aphids. Ants tend them and feed on the excess sap that the aphids secrete. But you may also see a small butterfly flitting about the aphids, sometimes landing on them and walking quickly over them. This is the wanderer, *Feniseca tarquinius*. It has a wingspan of one and a quarter inches and is colored with blotches of orange and brown. As it moves over the colonies of woolly aphids, it leaves an egg behind. In a few days this hatches into a larva that spins a tiny protective webbing over itself in among the aphids. Its mouth is different from that of other larvae because it can pierce the aphids and suck the juices from them. It then incorporates the empty woolly skins of its victims into its webbing and this effectively conceals its presence. In about eleven or twelve days it changes into a pupa, leaves the aphids, and

Harvester butterfly feeding on honeydew from aphids

pupates on a nearby twig or leaf of the alder. This is quite an unusual life cycle for the larval form of a butterfly, for most other butterfly larvae are herbivorous, not carnivorous.

LONGHORNED BEETLES AND SOLDIER BEETLES

Relationships

Longhorned beetles are a family (Cerambycidae) of insects in the order Coleoptera, or beetles. Their common name refers to their long, thick antennae, which are usually over half the length of their bodies. Many of their larvae bore holes in dead or living trees and feed on the wood. Adults are brightly colored and often feed at flowers.

Soldier beetles are a family (Cantharidae) of insects also in the order Coleoptera. They also have long antennae, usually a little less than half the length of their bodies, and they are oblong in general outline, like the longhorned beetles. Their larvae feed on other insects in the soil or under bark, and the adults are found at flowers feeding on pollen and nectar or eating aphids and other insects.

Life Cycle

Longhorned beetles overwinter as larvae in tunnels within wood. In spring they resume feeding on the wood and,

On goldenrod flowers: left, *soldier beetle;* right, *longhorned beetle. Life size*

Longhorned and soldier beetle habitat

when just about to pupate, burrow to the edge of the wood and create a little hollowed-out area in which to pupate. Pupation occurs in mid- to late summer. Adults emerge from pupae in late summer and fall and feed and mate on flowers. They lay eggs in or on the wood of host trees, and larvae hatch and feed in fall and then overwinter. Some longhorned larvae take two to three years to mature, instead of just one.

Soldier beetles overwinter as larvae, usually underground, and in late spring after resuming feeding they pupate in earthen cells. They emerge from pupation in summer, and the adults can be seen through fall. They mate, and the fertilized females lay clusters of eggs, usually under the soil. The larvae hatch from the eggs and begin feeding on small insects and other animals in the soil. They overwinter in this stage.

Highlights of the Life Cycle

One of the most accessible aspects of these insects' lives is their adult stage, when they are easily found feeding on flowers. You can compare the two types of beetles, identify them, and see all of their structures closely. They are among our most lovely beetles.

How to Find Longhorned and Soldier Beetles

Look along roadsides or in meadows for goldenrod and then look among the blossoms for long, thin beetles, three-quarters to one inch long, with antennae about half as long as their bodies. They are often colored with yellow, orange, black, and brown.

What You Can Observe

FOUR SPECIES OF BEETLES

Some of the most beautiful insects that visit goldenrod to feed on the nectar and pollen are the longhorned beetles and the soldier beetles. There is no easy and foolproof method of distinguishing between these two families of beetles without examining them closely. Both types are long, thin beetles with antennae about half the length of their bodies. Soldier beetles are closely related to fireflies and may remind you of them at first. The segment just behind the head that covers the back is usually flattened around the edges just as in the firefly. Longhorned beetles tend to be more cylindrical throughout their length, and they have a generally sturdier appearance.

Longhorned beetles get their names from their long antennae, which are often as long as or longer than their bodies. This distinguishes them from many other groups of beetles, though not, unfortunately, from the soldier beetles, for they also have long antennae. Although the soldier beetles look similar, their lives are very different. The adults feed on pollen and nectar but also eat other small insects, such as aphids. They lay eggs in the soil, and the larvae move about on the soil surface, feeding on the eggs and young of other insects. In fact, some species of soldier beetles are used to help control the larval forms of certain moths that are agricultural pests. Their mouths have an extra part that is fleshy and soft and may be used to help the adult insect lap up the nectar from the flowers.

These clues are not enough for the beginner to identify these beetles. However, there are four common species of beetles, two from each family, that habitually visit goldenrod flowers, and through pictures and descriptions they're easily recognized.

One common species of longhorned beetle that is quite well camouflaged on the flowers is the locust borer (*Me-*

Left to right: *locust borer, banded longhorned beetle, common soldier beetle, margined soldier beetle*

gacyllene robiniae), so named because its eggs are laid in the crevices of locust bark. Egg-laying usually takes place in the afternoon, while mating and feeding on goldenrod takes place in the morning. The larvae tunnel into the inner bark of the locust, hibernate there, and in spring excavate tunnels into the heartwood and sapwood. They mature in the tree and emerge as adults in late summer. They are long, thin beetles, basically black with yellow bands across the covers of their hind wings.

Another longhorned beetle is the banded longhorn (*Typocerus velutinus*). Its wing covers are dark red with lighter red bands crossing them. Its larvae commonly feed in decaying birch, but may also feed on a variety of other trees. Its life history is similar to that of the locust borer.

The two species of soldier beetle commonly found on goldenrod look very similar. The common soldier beetle (*Chauliognathus pennsylvanicus*) has yellow wing covers with black markings covering the back half of them, while the margined soldier beetle (*Chauliognathus marginatus*) has black markings that cover almost all of the wing covers.

All four beetles feed a great deal on goldenrod. Besides eating, the beetles also mate on the flowers. The flowers may function as a mating spot that is attractive to both sexes and enables them to locate suitable mates.

ACORN WEEVILS

Relationships

Acorn weevils are a genus (*Curculio*) of insects in the family Curculionidae, or weevils, which, in turn, is in the order Coleoptera, or beetles. Weevils are the largest family of all insects. There are over twenty-five hundred species of weevils in North America, four times the numbers of species of all birds in the same area. Most are less than one-quarter inch long, and many larvae feed and develop within seeds. All weevils have an extended mouth that projects out, a little like the trunk of an elephant. Acorn weevils have particularly long mouthparts.

Life Cycle

Acorn weevils overwinter as mature larvae buried a few inches beneath the soil. They may remain in this stage through one to three winters. Usually it is only one winter, and then in spring the larvae pupate in the soil. The adult

Acorns showing weevil exit holes. Life size

Acorn weevil habitat

weevils emerge in summer, and the fertilized females lay their eggs inside developing acorns. The larvae mature in the acorn, and, after the acorn has dropped to the ground, they leave it and burrow into the soil. They overwinter in that stage.

Highlights of the Life Cycle

You will rarely have a chance to find the adults, for they are up in the oak trees mating and laying eggs. But when the acorns fall, you can find the larvae in them, or, if the larvae have already gone, you can see much of the pattern of their lives by examining the acorn. There are other insects that can be found in the chewed-out portions of the acorns after the weevil has left.

How to Find Acorn Weevils

Go to areas of natural oak woods or to groups of oaks and gather some fallen acorns. Press them between your thumb and forefinger, and if there is some give to them, or if they have a hole in their shell about the size of a pinhead, then that acorn either has or has had acorn weevil larvae in it. Open up the acorn and examine the contents, and the following descriptions will help you know what you have found.

What You Can Observe

THE ADULT BEETLE

To understand what you find in the acorn, you need to be familiar with the habits of the adult beetle. The outstanding feature of this minute beetle is an extended portion of its face that forms a long beak, almost longer than the body. At the tip of this beak are small chewing mouthparts, and halfway back on the beak are the beetle's antennae, which bend in the middle. The basal portion folds into a groove along the beak, so that the antennae won't get in the way when the beetle uses its beak. The beak is used for drilling into acorns and feeding on the interior nutmeat. The female also uses her beak to drill through the shell of young

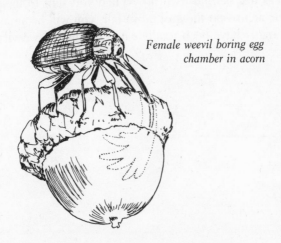

Female weevil boring egg chamber in acorn

acorns in early summer and make small chambers. She lays an egg in each one and then plugs the hole in the shell with an excreted fecal pellet, which dries and turns white. After hatching, the larvae feed on the developing acorn and, when mature, drill a hole through the acorn and enter the soil a few inches from where the acorn fell.

INSIDE THE ACORN

If you find an acorn that is soft but has no large exit hole, it means the larvae are still inside. When there is an obvious exit hole, only fecal pellets from the larval feeding will be found inside. There is one other insect you may find inside these acorns. It is a minute moth called the acorn moth that lays its egg in acorns that have been fed on by weevils. The egg hatches and the moth larva feeds on the remains of the acorn. It overwinters in the acorn, continues feeding in spring, and then pupates and emerges as an adult in midsummer. Either of these two larvae can be found in fall acorns, but they are easily distinguished, for the weevil larva has no legs, while the moth larva has three pairs of small legs just behind its head. When you start to look at all of the acorns on the ground in fall, you will be amazed at how many have been totally eaten out by the weevils.

MONARCH BUTTERFLY

Relationships

The monarch butterfly is a species (*Danaus plexippus*) of insect in the family Danaidae, or milkweed butterflies, which in turn is in the order Lepidoptera, or moths and butterflies. Some of our largest butterflies belong to this family. They feed on milkweed in the larval stage and gain some protection from predators because of this. Milkweed contains some toxins that the larvae are able to incorporate into their bodies without harm, and that make them distasteful to many predators.

Life Cycle

In fall, monarch butterflies migrate from eastern North America to overwinter in Mexico and Central America. In spring they migrate back, and some mate and lay eggs as they move north. Eggs are placed singly on the leaves of

Monarch butterfly habitat

Monarch butterfly. Life size

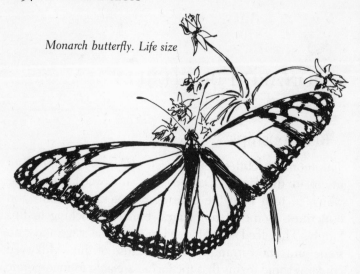

milkweed. They hatch in about four days, and the larvae feed on milkweed leaves. Over the next ten days, the larva molts four times and is then ready to pupate. The pupa is attached under a leaf or onto some other solid object. Pupation lasts about twelve days. There may be three to four broods, and the adults of the last brood migrate south in the fall and overwinter as adults.

Highlights of the Life Cycle

It is hard to choose a highlight in the life of the monarch, for so much of its life can be observed and so much of it is both beautiful and fascinating. Watching the caterpillars transform to pupae and then to adults is easily done by bringing them inside and feeding them milkweed leaves. Look for the mating behavior of adults as they move about and feed on milkweed flowers, and also for their strong and directed flight as they migrate south in the fall.

How to Find Monarch Butterflies

Look along roadsides or in fields where there are flowers and watch for a large butterfly (wingspan three and a half to four inches) that is orange with black stripes, flying strongly just above the flowers. The only other butterfly like it is the viceroy, with which it is easily confused. The best way to distinguish the two is to look at their hind wings. Both have orange wings with black veins and a black rim, but the viceroy has an additional black marking cutting across the veins in the middle of the wing. See illustration in winter section under viceroy butterfly.

What You Can Observe

COURTSHIP BEHAVIOR

During summer you may have a chance to see some of the courtship activities of monarchs, especially in meadows where the adults are feeding on flowers. The male tends to perch on exposed sunny places, such as the tips of weeds or branches, usually between five and fifteen feet high. As other large butterflies come by, it flies out to them to determine if they are female monarchs. It will fly after viceroys and swallowtails as well as other monarchs. If the butterfly is a female monarch, the male usually flies behind her for a short distance and bumps into the tip of her abdomen. Following this, the male and female engage in a fast flight that looks like a speedy and erratic chase. This

Left: *viceroy butterfly*; right: *monarch butterfly*

may last up to a minute, cover hundreds of yards, and go up to a hundred feet high, or it may be brief. Following this, the male tends to grab on to the female from above and hold his wings straight out and still, while the female flutters some and the two glide to the ground. Usually, while hidden among weeds, the male attaches his abdomen to the female's abdomen and then carries her off, flying, while her legs are folded beneath her and her wings are closed. They fly to dense vegetation where they remain from two to fourteen hours while the packet of sperm is transferred from the male to the female. Rather than ever seeing this complete set of behavior, you are more likely to see only parts of it, since courtship is not always successful and vegetation may obscure your view.

MIGRATION BEHAVIOR

Even casual observation of monarch butterflies will reveal some interesting changes in their behavior over the course of the summer. Watch the butterflies as they move between flowers during feeding. In early summer they will wander in any direction to the next flower, but in late summer you will notice that they most often move in a decidedly southern direction when choosing the next flower. This is be-

Larva and chrysalis of monarch butterfly

cause the adult monarchs of late summer migrate south for hundreds of miles to places where they spend the winter.

When these migrating butterflies encounter obstacles, such as forests, buildings, ridges, or mountains, they tend to just fly higher, right over them. This is why even on the tops of smaller mountains you can find monarchs flitting by in the fall.

During the nights they may rest singly, in small groups, or in large clusters. The resting places for the night are usually trees with narrow leaves or narrow projections that afford a good place for the butterflies to hold on. Favored trees are willows, maples, and pines. Single butterflies may roost anywhere. Small groups of butterflies may roost at the edges of rivers, ravines, and meadows. The large roosts of hundreds to thousands of butterflies form at the edges of large bodies of water, usually at the southern tip of projecting points or peninsulas of land. These include areas along the Atlantic coast, the Gulf Coast, and the Great Lakes. If these spots have a number of good trees and are fairly protected, they may be used year after year by the migrants as roosting spots. The final destination of migrating monarchs from the East is the mountains of southern Mexico, where they overwinter in aggregations of millions in special areas of only twenty or thirty acres. Monarchs in the West move south along the coast and winter along the California coast between San Francisco and Los Angeles. They use groves of Monterey pine and eucalyptus as roosting sites.

In spring, the first monarchs appear in the South in March and April, in the central states in April and May, and in the North in May and June. As they move north, some stop to lay eggs on the young milkweed plants in the southern areas, while others wait until they reach the North before laying eggs.

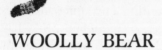

WOOLLY BEAR

Relationships

The woolly bear is a species (*Isia isabella*) of insect in the family Arctiidae, or tiger moths, which in turn is in the order Lepidoptera, or moths and butterflies. Tiger moths are best known, and often named, for their larval stage, during which they are large, active, furry caterpillars. In fact, if you see a large caterpillar with a dense covering of long stiff hairs, there is a very good chance it is a tiger moth caterpillar, for very few other caterpillars have this

Woolly bear habitat

Woolly bear crossing path. Life size

quality. Adults are called tiger moths because many have bold patterns of stripes on their wings.

Life Cycle

The woolly bear overwinters as a larva curled up under loose bark or in some other protected place. In spring it resumes eating leaves and soon pupates in a cocoon that contains most of the caterpillar's hairs held together with silk. Pupation lasts about two weeks. Adults are active in early summer, and fertilized females lay eggs in clusters on a variety of plants that the caterpillars eat. Eggs hatch in four to five days and caterpillars initially feed together. After the first few molts, they feed separately. In three to four weeks, larvae have undergone six molts and crawl about looking for a protected place to pupate, such as under a board or log. Adults emerge from the pupae, lay eggs, and the larvae from this second generation overwinter.

Highlights of the Life Cycle

The larva is the most conspicuous form of this insect, especially those of the fall generation, for they spend a great deal of time crawling about in search of good overwintering spots, places where they will be protected from predators and sudden changes in temperature. They are often seen crossing roads and paths and can be picked up without harm to either you or the caterpillar.

How to Find Woolly Bears

In September and October, keep on the lookout for furry red-brown and black caterpillars crossing roads and paths. They are about one and one-half inches long, with a dense covering of quarter-inch-long hairs all over them, and the hairs are in three bands of color, black at each end and red-brown in the middle. They curl into a ball when disturbed.

What You Can Observe

STRUCTURE AND BEHAVIOR OF TIGER MOTH CATERPILLARS

The larvae of most moths and butterflies are difficult to identify, not only because there are no adequate identifi-

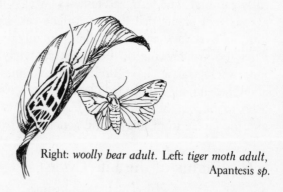

Right: *woolly bear adult*. Left: *tiger moth adult*, Apantesis *sp.*

cation guides, but also because there is a great diversity of forms both between species and within the different stages of the same species. But the woolly bear and, in fact, most tiger moth larvae have made things easier. They are the only caterpillars common in fall that are large and covered with long hairs.

Their behavior is also a clue to their identity. After finishing their feeding for the season, they actively crawl about the ground looking for a place to spend the winter. This is when we most often cross their paths. They can be difficult to pick up because either their long hairs come off in your hand, or the quality of their stiffness makes them give way when touched. Some species, such as the woolly bear, tend to curl into a tight ball when disturbed. They are incredibly "slippery" in this position and slide right off your hand as if it were a surface of ice. These are probably all protective measures evolved in reaction to various bird and mammal predators.

The woolly bear is the species most commonly seen in fall. It seems to race across sidewalks and dirt paths, and indeed it may be fast for a caterpillar. Its actual speed is about four feet per minute (approximately .05 miles per hour). It will hibernate during the winter under a rock or log in the larval stage and then in spring emerge to feed for a while before forming a cocoon and pupating.

RECOGNIZING OTHER SPECIES OF TIGER MOTH CATERPILLARS

Here are a number of other tiger moth caterpillars to look for. Some can be located and recognized by their specific host plants such as:

milkweed tiger moth: found on milkweed; even-length yellow hairs in the center, longer black and white hairs at each end.

dogbane tiger moth: found on dogbane; light tan hairs all over, not as heavily furred as other tiger moth caterpillars.

Curled up: *woolly bear*; top: *yellow woolly bear*; bottom:
milkweed tiger moth caterpillar

great leopard moth: found on plantain; all black hairs, red
bands between body segments seen through the hairs.

Others are more general in their feeding habits but still
can be recognized because they're common in fall, they
curl into a ball, and have long hairs on their bodies.

hickory tiger moth: found on hickories, walnuts, and other nut
trees. Hairs are mixed black and white.
spotted tiger moth: often on poplar or maple, hairs mostly
white; longer hairs at each end of their bodies.
yellow woolly bear: found on a variety of trees. A little like a
light-colored version of the woolly bear, except that its hairs
are of unequal lengths, and although evenly colored, it
varies with individuals from light tan to deep red-brown.

FALL WEBWORM

Relationships

The fall webworm is a species (*Hyphantria cunea*) of insect in the family Arctiidae, or tiger moths, which in turn is in the order Lepidoptera, or moths and butterflies. It is best known for its larval stage, which, unlike most other species of tiger moth, builds a communal web tent over the leaves that it feeds on. The adults are stout-bodied moths with wingspans of a little over two inches. The woolly bear is

Fall webworm habitat

Fall webworm and eaten leaf. Life size

another member of the same family with habits more typical of other arctiids.

Life Cycle

Fall webworms overwinter in the pupal stage encased in a light cocoon placed under loose bark or debris on the ground. Adults emerge in late spring and early summer, and the fertilized females lay eggs on the leaves of their various food plants. The eggs are placed in masses and covered with white hairs from the female's body. Eggs hatch in about a week, and larvae begin to construct communal web tents around the leaves they are eating. Caterpillars undergo about six molts in four to six weeks, then leave the web nest and singly look for places to pupate. In the south there are two broods, but in northern areas there is only one.

Highlights of the Life Cycle

The web tents and the habits of the larvae within them are by far the most accessible part of the life cycle of these insects. The web nests are most conspicuous in fall, and even if you find them after the caterpillars have left, there are many interesting things to discover within them.

How to Find Fall Webworms

Look along roadsides and woods borders for masses of webbing enclosing leaves at the tips of branches. The webbing may extend a foot or more along the branch. They feed only on deciduous trees, and among the more commonly chosen are ash, cherry, willow, and apple.

What You Can Observe

THE WEB NEST

Once the young caterpillars hatch, they immediately begin to build the web. Building occurs mostly at night. The nest increases in size until it may be up to three feet long and encompass several branches.

Fall webworm caterpillars seem to start building just about when the tent caterpillars are finishing their nests, and the nests of the two are similar. One simple thing that distinguishes them is that tent caterpillar nests are built in

Fall webworm web nest

the forks of branches, while fall webworm nests are built at the tips. This difference in placement reflects the different uses of the nests. Tent caterpillars leave their nests to feed on leaves, whereas the fall webworms build their nests around the leaves they feed on. Hence, you rarely see fall webworm caterpillars outside their nests. Only just before the caterpillars are about to leave and pupate do they crawl around outside the nest and feed, unprotected, on leaves.

INSIDE THE NEST

If you poke through either active or abandoned nests, you can find a number of interesting things. One is the shed skins of the caterpillars, shed as they go through the six molts in their development. Over a quarter of the caterpillar's life is spent in the process of shedding skins, or in quiet periods after shedding when its new skin is hardening. Inside the nest you will also find all of the caterpillars' droppings. The caterpillars stay mostly in the nest during the day, but will leave and rest on the outside if the temperature inside becomes too hot. After molting five times, the mature webworm leaves the nest and forms a cocoon, placed in bark crevices or under leaves on the ground, in which to pupate and spend the winter.

While looking through the webbing, you may discover a caterpillar that doesn't move at all. This caterpillar may have been parasitized by a small ichneumon wasp, *Hyposoter pilosulus*. This wasp enters the webbing and lays its eggs in the young webworm larva. The eggs hatch and feed inside the larva. As they get larger, their feeding kills the larva. They spin a cocoon inside the larva and pupate there. In fall they emerge as adults and spend the winter in that stage.

Adult fall webworm

PREDATORS AROUND THE NEST

The nest would seem like perfect protection against predators, and in fact it may protect the caterpillars significantly,
especially in their earlier moltings. Most attacks by predators start around the fourth and fifth moltings of the caterpillars. Two of the main predators on webworm nests are
birds and hornets. In general, once a nest is discovered by
one of these predators, it is continually revisited, sometimes
until the whole colony of caterpillars is wiped out. Yellow
warblers have been observed to feed frequently on the
webworms, and two species of hornets also eat them. One
species, *Vespula arenaria*, has the habit of diving down on
a caterpillar that appears near the surface of the webbing,
chewing away at it for a while, and then carrying it away
to its own nest to feed to its developing larvae. Another
habit is seen in V. *maculifrons*, which enters the nest
webbing, catches a larva and stings it several times. It then
chews it up into a ball, which it carries away to its nest.
Keep on the lookout for wasps flying around webworm
nests and you may see one of these behaviors.

HOUSEFLIES

Relationships

Houseflies is a term loosely applied to a family (Muscidae) of insects in the order Diptera, or flies. They are small, stocky flies with bristles over their bodies and often a series of black lines along the back of the thorax. Many species lay their eggs in decaying matter or dung. Some species are typically found in our houses and other buildings.

Life Cycle

Three species, the housefly, the face fly, and the lesser housefly, are mentioned here, and their life cycles are

Housefly habitat

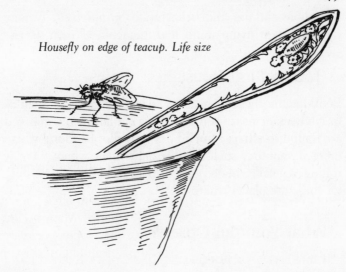

Housefly on edge of teacup. Life size

roughly similar. These flies are known to overwinter in the adult stage, but may possibly overwinter in other stages as well. They overwinter in protected places, such as within the walls of houses, in attics, or under loose bark. In spring they become active, and fertilized females lay eggs in four or five batches of over a hundred each. Eggs are laid in decaying material, such as garbage, rotting flesh, or dung. Eggs hatch within a day, and the larvae, or maggots, mature in five to seven days. They pupate for another five to seven days and then emerge as adults. At least in the case of the housefly, the adults fly about and feed for another two weeks before mating. There may be ten or more broods in a year, depending on the warmth of the climate.

Highlights of the Life Cycle

These flies are always conspicuous around and in buildings, at least in the adult phase. The adult flies have some incredible abilities and behavior to be observed. Among these are their flying ability and the acuity of their senses.

How to Find Houseflies

Look for the flies where there is food or garbage, such as in kitchens or near trash cans. Have a fall picnic and you are bound to attract some. They may also be found where there is manure, such as near a barn, or where there is rotting vegetable matter, such as next to a compost pile, for these are places where they lay eggs.

What You Can Observe

THREE TYPES OF HOUSEFLIES

The average person probably has done more casual observing of houseflies than of any other insect. And yet, since flies in our houses are looked upon mostly with disgust, the observations were probably not terribly objective or scientific. You may recall three common housefly experiences. One is in late winter and early spring, when you find tens of flies buzzing against attic windows trying to get out. Another occurs in early summer when you see a few smaller flies hovering about a light fixture in the center of a room, making sharp angled turns and occasionally landing. And then there are the larger flies in late summer that are always landing on the kitchen counter and rubbing their legs together, or jumping on another fly nearby. Actually, each of these is a different type of fly. But they are all in the family Muscidae, which, in general, includes stout-bodied, dark-colored flies covered with bristly hairs.

The first one, found in such abundance in third-floor rooms or attics in spring or warmer days of winter, is actually not a fly that spends the summer in the house. It is likely to be *Musca autumnalis*, or the face fly. This fly was introduced to North America in 1952 and is already widespread. Its summer life is spent almost entirely in open pastures where there are cattle. Here these flies lay their eggs in the cow manure, where the young larvae develop. They spend the whole warm part of the year away from houses; in fact, the flies even stay outside when the cattle go into the barn at night. The flies are called face flies because they often land around the face of cattle, feeding on the exudations from the animals' eyes or noses. When the weather becomes colder, the flies leave the pastures and the cows and spend time sitting on the sides of buildings where the later afternoon sun will warm them. Then, as the sun goes down, the flies crawl into protected crevices in the buildings. When it is finally too cold for them to remain outside at all, they move inside a house.

The next fly seen regularly inside of the house in spring is the lesser housefly, *Fannia cannicularis*. This fly gets started earlier than the housefly and has slightly different habits. It is often seen in the center of rooms under a light fixture or under a pull string to a light. It stays in the center of the room and flies about in a series of short, straight flights interrupted by sharp-angled turns that eventually bring it around in a circle or, rather, some kind of polygon. Occasionally there will be a number of flies doing this at once. Sometimes one will land on another and the two will seemingly grapple for a second and then separate. Most of the flies in these groups are males, and it is presently believed that this hovering behavior is a type of swarming, with the light fixture in the center of the room being the swarm marker, and that when females enter the area, males dive upon them and mate with them. Males also occasionally land on other males, possibly mistaking them for females or aggressively attacking them. These flies hover more than the housefly and they are slightly smaller. They, of course, cannot be small houseflies that will get bigger in late summer, since all flies go through complete metamorphosis and, once adults, never grow larger.

The larger flies usually seen in July and August are the common housefly, *Musca domestica*. This is the fly you have probably observed most, since it is large and makes human homes its constant abode. These flies can also be

Housefly larva and pupa

seen mating, but they don't swarm. The male approaches the female, possibly attracted by an odor that she releases. When within an inch of her, he jumps onto her back, vibrating his wings with an audible buzz. The female puts her wings out to the sides and hooks her second pair of legs over them. The males then takes her first pair of legs in his first pair of legs and moves them up and down. Meanwhile, he moves backward on top of her and copulates. Like the lesser housefly, the male of the common housefly will sometimes jump on top of other males and they will grapple and then separate. The female houseflies can be distinguished from the male houseflies by the fact that their eyes are spaced farther apart; the male's eyes practically meet in the middle of his forehead.

FLY MOUTHS AND FEET

None of these flies bites. In fact, they can't bite: their mouths are not designed for it. So if you are bitten, don't blame it on one of these flies. A fly's mouth consists of a little vacuum-cleaner-like projection that has two roughened flaps at its tip to help break up surfaces and lap up juices. These flies only eat exposed liquid foods, but in some cases they will regurgitate fluids that help predigest the food and then they lap it up. Flies taste with their feet and not with their mouths, so as they walk across a surface and their feet contact liquid with food content, their mouthparts automatically descend and they feed.

The feet of flies are remarkable for their ability to hold on to surfaces, such as glass or ceilings. The tip of each foot has a pair of small claws that help it attach to rough surfaces. In addition to this are two tiny pads below the claws. Each is covered with minute hairs that can exude a sticky substance at their tips.

FLIES ON WINDOWS

Finally, there is a peculiar event that you may have the chance to observe in fall. Look upon your windows for a housefly that seems to be stuck there and remains motionless even when you move your hand above it. If you look closer, you may see bits of a white powdery substance on the surface below the fly. This is the result of a fungus that enters the system of the fly as a spore. The spore grows within the adult fly. When it is about to release new spores, it has a way of attaching its host to a spot. It then releases the spores from the fly's body. All that will be left of the fly in these cases is an empty outer skeleton.

SAWFLIES

Relationships

Sawfly is a general term referring to at least seven families of insects in the order Hymenoptera, or bees, wasps, and ants. They are believed to be one of the more primitive members of this order. Adult sawflies look a little like bees

Sawfly larvae on red pine needles. Life size

Sawfly larva habitat

and wasps, except on sawflies there is a broad attachment joining the thorax and abdomen, whereas in bees, wasps, and ants the attachment is narrowed to a threadlike waist. Sawfly larvae are very similar in appearance to the caterpillars of moths and butterflies, and they feed on leaves and within stems as well. Some of the more common families of sawflies are: common sawflies (Tenthredinidae), cimbicid sawflies (Cimbicidae), conifer sawflies (Diprionidae), and stem sawflies (Cephidae).

Life Cycle

Since there are so many types of sawflies, this is necessarily only a general description of their life cycle. Features of certain species are described in later sections of this entry. Sawflies overwinter most often as larvae in cocoons that are either attached to twigs, within twigs, or in or on the ground. Although some transform to pupae in the cocoons before winter, most wait until spring and then pupate. Length of pupation is unknown for most species, but it is probably about one to two weeks. Adults emerge in late spring and early summer, and mate and lay eggs on or in plants that the larvae will feed on. The eggs hatch in about a week and the caterpillarlike larvae feed externally on foliage, or within stems on the wood. When mature, they form cocoons in which to pupate, and overwinter as larvae or pupae in the cocoons. There may be one to three broods per year, depending on both the species and the length of the warm season.

Highlights of the Life Cycle

The highlight of the life cycle depends a great deal on which family of sawflies you find. In the conifer, cimbicid, and common sawflies, you are most likely to find the larvae and be able to observe how their structure and behavior are different from that of caterpillars. You may also find their cocoons. In the case of the stem sawflies, you are more likely to see evidence of their feeding in distorted parts of their food plants and find clues to their behavior from signs on the plant.

How to Find Sawflies

Although sawfly larvae feed on a variety of plants, such as fruit trees, brambles, dock, and violet, I find that those feeding on pines or other evergreens are the easiest to locate in fall. At the tips of pine branches, look for caterpillarlike larvae, each about an inch long, feeding together on the needles. Some common species are green or black with red and yellow lines or dots. When disturbed, they pick up their heads and form their bodies into an S curve. While looking for the larvae, you may come across one of the cocoons. They are oblong, one-quarter inch to one inch long, and made of tough, brown, parchmentlike material. They are fastened lengthwise to the twigs, and sometimes one end is neatly cut off, making the cocoon look like a little cup.

What You Can Observe

THE LARVAE

Sawflies are one of the best-kept secrets in the world of insect study. You can read book after book on insects, many of them quite detailed, and still know nothing about the many species of sawflies common in our woods. Saw-flies are in the same order as bees, wasps, and ants —

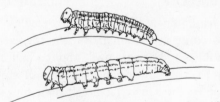

Larvae: sawfly above, moth or butterfly below

Hymenoptera — but amazingly, their larvae are very similar to the caterpillars of moths and butterflies. They even have many of the same habits: feeding on leaves, boring into stems, creating galls (deformations of plant tissues in which the insects live and feed).

The easiest way to tell if you have a caterpillar or sawfly larva is to look at the legs. Both insects have three pairs of hardened legs just behind the head. After this are pairs of

Sawfly cocoons attached to pine needles

softer, leglike projections called prolegs. The sawflies have six or more of these; the caterpillars have five or less. Sawfly larvae have two other distinctive features: they curl the tips of their abdomens around the object on which they are feeding, and they curl their heads up and back when disturbed.

THE COCOONS

While searching among the pine needles, you are also likely to come across small, brown cocoons stuck to the twigs or foliage. They are oval, about a half inch long, and made from extremely tough, parchmentlike material. They are made by the sawfly larvae. If their tops are neatly cut off, the adult sawfly has already emerged. If you find just pinholes in the cocoons, it means the larvae have been parasitized by an ichneumon; these are the exit holes of the adult parasites. If you find no holes in the cocoon, the sawfly is probably still developing inside.

Sawflies make their cocoons either attached to twigs or among leaf litter on the ground. Obviously we have the best chance of finding those attached to twigs. Along with pines, a good place to look for the cocoons is on the twigs of shrubs beneath pines, for the larvae often drop down to make their cocoons lower. These are most easily spotted in winter when there are no leaves on the shrubs. One particularly large sawfly cocoon is found in deciduous woods. The cocoon is an inch long and belongs to the American sawfly (*Cimbex americana*), the adult of which looks like a large hornet.

THE ADULTS

As mentioned earlier, adult sawflies look a little like bees, wasps, and ants, except for the difference in the way the abdomen and thorax are joined. The adult sawflies are generally active in spring and summer, when the female lays her eggs in the leaves of the plants on which the larvae will feed. Her ovipositor is slightly sawlike, giving the insect its common name. Look on the pine needles for evidence of the egg-laying sites. They usually appear as rows of little brown marks along the edge of the needle.

A STEM SAWFLY

Evidence of another type of sawfly can be found on willow shrubs; look for the tip of a willow shoot that is dead and curled over. This is the work of common willow shoot sawfly. The female chooses fresh willow branch tips and, a few inches from the tip of the stem, inserts her ovipositor ten or more times in a circle around the shoot. She then lays an egg in the stem above where she made the incisions. The shoot soon wilts and dies and the larva bores out to the tip of the shoot, then turns around and bores in a spiral fashion down into the main stem and even into the root. When it is ready to pupate, it crawls into the aboveground portion of the stem and eats a hole about a quarter of an inch in diameter, reaching just to the edge of the bark but not going all the way through. It then pulls back and pupates in a cocoon inside the stem, emerging the next spring. In winter it is usually still a larva inside the cocoon, and pupation generally takes place in the spring. When you have found one of these damaged shoots, you can look for the holes where the female killed the twig; you can break off some of the dead twig and look for the larval gallery; and you can look for evidence of the emergence hole lower in the stem.

Also on willows is evidence of another group of sawflies, the ones that make galls. Many of the galls on willow leaves that are hard, globular, and about a half inch across are made by sawflies. These are in the family Tenthredinidae.

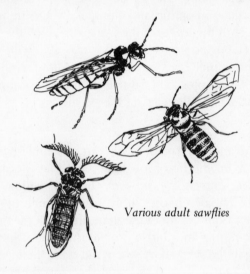

Various adult sawflies

EVOLUTION IN HYMENOPTERA

Sawflies are thought to be a more primitive type of Hymenoptera because they have no stinger and their larvae feed on plants. It is thought that wasps were plant feeders first, and that the females developed long ovipositors so they could insert their eggs into plants. Some then developed chemicals that were excreted with the eggs, and these made galls. Soon, some wasps evolved ways to lay their eggs in other insects, and the chemicals had the effect of paralyzing the host. All of these insects lived nonsocial lives. When the insect societies began, the ovipositor and its chemicals were used for defensive stinging and for egg-laying. Bees are believed to be the most advanced, having gone back to feeding on plants and having kept the stinger as defense in the social species. In the sawflies, we are seeing an example of the earlier stages of this evolution.

HORNETS AND YELLOW JACKETS

Relationships

Hornets and yellow jackets are members of a family (Vespidae) of insects that are in the order Hymenoptera, or bees, wasps, and ants. It is a large family that also includes potter wasps and paper wasps. Many members of this family are social, and these make nests of papery material. Wasps in this family can be recognized by the way they hold their wings when at rest — folded and slightly out to each side of the body, rather than over the back as in other families of wasps. The terms *hornet* and *yellow jacket* are common names loosely applied to many species of vespids that build covered paper nests and that have yellow and black mark-

Hornet and yellow jacket habitat

ings on their abdomens. There is no strict division between the two except that, in general, hornets build nests aboveground and yellow jackets build them underground.

Life Cycle

Hornets and yellow jackets overwinter as fertilized queens. In spring the queens become active, gather nesting material, and start a small nest either underground or aboveground, attached to plants or buildings. After making a few hexagonal cells and a covering around them, the queen

Yellow jacket on apple core. Life size

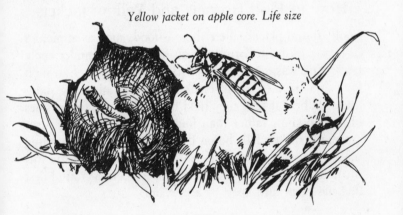

lays an egg in each cell. The eggs hatch in a week and the queen feeds the larvae small bits of prey. She continues this for the next ten to twelve days. Then the larvae pupate in their cells another twelve days. After emerging as adults, these sterile females start working for the queen, collecting food, enlarging the nest, and tending the new young. In late summer the queen lays eggs that develop into males and fertile females. These mate, and the fertilized females overwinter. Eggs stop being laid and the hive breaks up after the males and fertile females leave.

Highlights of the Life Cycle

The adult insects are active from late spring until mid-fall, but as the hive grows they become more numerous, so by fall there are many more wasps in the air. Also, their social structure begins to break down and individuals are more commonly seen feeding on fallen fruits or other foods. As cold weather approaches, the nests are often attacked by mammals, which eat the cold and sluggish insects inside. If no live wasps are still in these nests, you have a great chance to examine their structure.

How to Find Hornets and Yellow Jackets

Look around places where there is food, such as at picnics or near park trash barrels, or on fallen apples under apple trees. The wasps are one-half to three-quarters of an inch long and have bands of black and yellow on their abdomens. When at rest, they hold their wings out to the sides of their bodies, rather than folded over their backs. Also, especially in late fall, be on the lookout for bits of nesting material on the ground. If you find some, check the immediate vicinity for a possible above- or below-ground nest that has been torn apart by a predator.

What You Can Observe

FEEDING HABITS

The fact that you see more hornets and yellow jackets feeding at your picnics and on the fallen fruit of orchards in fall is not just because you are having more picnics or because of the availability of the fruit. It also marks a change in the behavior of these wasps, associated with a turning point in the growth of their colonies. The wasps you see are most likely the workers (sterile females) of the colony who, through spring and summer, concentrated on making the paper nests and on feeding the young. But in late summer these workers stop feeding the young, and in some cases even break open the larval cells and feed on the larvae. They also spend less time at the hive and more time on their own, feeding primarily on sweet liquids. This change in behavior coincides with the final maturing of the male and fertile female wasps needed to carry the hive through winter, and the cessation of egg-laying by the queen. No one knows exactly why the change occurs.

Earlier in summer, you have a chance to see the workers hunt for prey to feed the larvae. Usually prey is other insects, such as caterpillars or flies, which supply the needed protein for the young. When hunting for flies, the wasps remain in areas such as barns or the sunny sides of buildings. You may see the hornets flying against a wall with a crash as they catch or miss a fly. They may also seem to be buzzing around you or at you, when, in fact, they may be trying to catch other insects in the vicinity.

THE NESTS

The nests of hornets and yellow jackets are made of gray, paperlike material. They are composed of horizontal layers of cells surrounded by several layers of paper covering. They may be built aboveground surrounding weed stems, or suspended from the branches of shrubs or trees. And they may be built underground in an old animal burrow or natural crevice. They are not built all at once but are the result of gradual growth. In spring they are started by the queen with a single layer of cells and a few layers of paper around it. The nest gets bigger as the worker wasps eat away the inner layers of the shell and add this material along with new material to the outside.

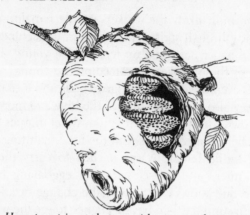

Hornet nest in apple tree, with cutaway showing interior

The nests become more obvious in fall, the aboveground ones because they are at their largest and are now in leafless trees and shrubs, and the underground ones because they are attacked by mammals, such as raccoons and skunks, who dig them up to eat the larvae made sluggish by the cold. As they do so, they scatter bits of the nest. The nests usually can be safely examined after the first few weeks of frost have killed the inhabitants. No wasps overwinter in the hives, although you may find many other smaller insects doing so.

If you can see inside the nest, look at a layer of cells. You will see some with an extension of about a quarter of an inch at their open ends. These extended cells will tend to be in the center of the nest. They have been used to raise two larvae to maturity. The cell is extended because when the first larva pupates, it excretes its waste material in the top of the cell. This makes the cell just that much too short for the next larva. Instead of cleaning it out, the cell is lengthened. At the edge of the layer you will see some cells only partially built. A little farther in, there will possibly be some cells with eggs or developing larvae now dead. Still others may be capped over and contain pupae, also dead. The queens are away from the nest, overwintering in protected crevices.

PAPER WASPS

Relationships

See Paper Wasps in the spring section.

Life Cycle

See Paper Wasps in the spring section.

Highlights of the Life Cycle

During the spring, it is primarily the behavior of the queens that is evident. In fall, it is the behavior of males and fertile females that is most outstanding, as they fly about gathering food from flowers, and as they gather in sunny areas to mate.

How to Find Paper Wasps

These are slender wasps about an inch long. Both their bodies and wings are dark brown; sometimes you can see

Paper wasp on goldenrod flowers. Life size

Paper wasp fall habitat

dark, reddish patches on the sides of their abdomens. In fall you may find them feeding on goldenrod or involved in mating activities on the sunny sides of vertical structures, such as telephone poles, tree trunks, or buildings. Nests are a single layer of paper cells about three or four inches in diameter, often hanging from under the eaves of buildings.

What You Can Observe

DISTINGUISHING MALES FROM FEMALES

Up until fall, the whole paper wasp society has been female, consisting of the queen and her sterile female workers. But in late summer and fall, males and fertile females (future queens) are raised by the queen and workers. Soon after the mature males begin to emerge from pupation, the rigid social structure of the nest begins to alter. The queen stops laying eggs and soon leaves. The workers do less feeding of the young.

At this time you will begin to see many of the males feeding on the pollen of flowers, especially goldenrod. The males are fun to watch close up, and you can do so since

Paper wasp adults: female on left, male on right

they have no stinger and cannot hurt you. The old joke says that if you pick up a wasp and it stings you, you know it is a female; but there are better ways to distinguish males from females. The males have a whitish to light yellow face, whereas the females' face is all dark brown. They are sluggish and when touched or pushed around, tend to just crawl away. One of my braver naturalist friends even picks up male paper wasps in his hands. I can easily recognize the males, but I still have qualms about picking them up.

MATING BEHAVIOR

On warm fall days you may also see mating going on, and if you are near the mating site, you might think at first that the wasps are attacking you or are at least angry and to be avoided. Mating sites are most often the sunny sides of tall, prominent objects, such as the side of a barn or building, or even the side of a telephone pole. Males often have a perch from which they dart at females, and this frequently is near spots where females would go to hibernate for the winter, such as crevices in buildings. The males assume their characteristic aggressive posture — body raised high and wings held out — and watch all passing insects. Usually after a male catches a female, he plummets to the ground and mates with her, then returns to the same perch. The attitude of the males while on these perches and during mating is very different from their otherwise slow and passive behavior.

After the mating period, both males and females crawl into more protected areas, where they spend the winter. The males, however, do not live through the winter, possibly because they do not have as much fat stored in their bodies. In spring the females emerge and start colonies as described in the spring section.

HONEYBEES

Relationships

Honeybees are members of a family (Apidae) of insects in the order Hymenoptera, or bees, wasps, and ants. This is a large family which also includes carpenter bees, bumblebees, and hundreds of species of solitary bees. The honeybee is not native to North America but was brought here by European settlers.

Honeybee habitat

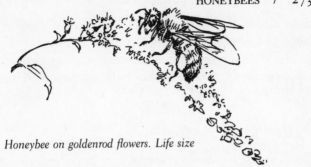

Honeybee on goldenrod flowers. Life size

Bees, unlike wasps, feed entirely on pollen and nectar. They are strict vegetarians, while most wasps feed primarily on other insects.

Life Cycle

Honeybees overwinter as adults. The typical hive includes a queen, workers (or sterile females), and drones (or males). In spring the queen starts to lay eggs and the workers tend them. Eggs hatch in three days. The larvae are fed by the workers and mature in about six days, and then the wax cell containing the larva is capped over by the workers, and pupation continues for twelve to fourteen days. In summer, workers live about six weeks and drones about eight weeks. Sometime in spring or summer, new queens develop in the hive, from normal worker larvae that are fed special food. These leave the hive along with some of the workers. They mate with drones and then take their workers to a new area and start another hive.

Highlights of the Life Cycle

Although there are many fascinating features of bee life in the hive, these are impossible to see without a special hive. Fortunately, there are just as many amazing things occurring outside the hive, and you can observe these at any patch of flowers. You can watch bees all summer as well, but in fall they are particularly active on goldenrod and other composite flowers as they build up their food reserves for the winter.

How to Find Honeybees

Go to any patch of flowers (the bees favor large groups of flowers, for they can keep returning to the same spot) and look for golden-colored bees covered with fine, short hairs. They are one-half to three-quarters of an inch long and will be busy gathering pollen and nectar. The bees are most active on warm, sunny days.

What You Can Observe

BEHAVIOR AT FLOWERS

Most popular works or even semitechnical works on honeybees go to great lengths to describe the fascinating social structure of these bees, and even describe how they communicate the whereabouts of food resources outside the hive. This is marvelous material, but for the average naturalist there is one drawback: you cannot see any of it unless you can get inside the hive or have an observation hive with glass walls. That leaves us with a lot of "book learning" about honeybees and very little learning from observation, which is a real loss with such interesting creatures. The best place to observe honeybees outside the hive is at flowers, and what the books fail to describe is the wonderful game that is going on between the bees and the flowers, many elements of which you can see and explore for yourself.

There are two sides to the "game": the flower's and the

bee's. Basically, flowers need to have the pollen, or male part, brought from one flower onto the receptive pistal, or female part, of a different flower. This results in cross pollination, a slight mixing of genetic traits, which provides a species of flower with more ways of surviving changes in the environment. The idea is for the pollen to be picked up on or by the pollinator and be rubbed or dropped off at the next flower. Some species of flowers have their pollen blown on the wind, but this necessitates a great deal of pollen production to ensure that enough females will get fertilized. Other species rely on animals to carry the pollen from flower to flower. But this has some drawbacks too, for animals don't voluntarily spend their energy helping out flowers. In fact, they probably wouldn't visit flowers at all if flowers didn't offer food in the form of nectar and pollen. Flowers have hundreds of different strategies designed to lure insect pollinators, but that is the topic for another book. Here we are concerned with the pollinator's strategy for collecting its own food and, while so doing, pollinating the flowers.

What honeybees are "trying" to do when they gather food is to get the greatest amount of food by expending the least amount of energy. (I say "trying," for these are not things bees are thinking about; rather these are strategies that are molded through evolution. The most efficient of them are passed on to the greatest number of offspring.) The best place to watch bees forage is at a large patch of flowers where there is more than one type of flower and preferably more than one color, such as among goldenrods and asters. Watch a bee arrive at the flowers and see which types it visits. The chances are great that it will collect nectar or pollen from only one type of flower. When the bee, after going from flower to flower, has gathered a full load of food in special spots on its hind legs, it will return to the hive. You may also see other individual bees repeatedly visiting other types of flower in the same area.

This "flower-constancy" strategy has two advantages for foraging bees. A bee has to learn how and where the nectar and pollen are located on a species of flower. With practice, the bee gets better at locating these and so can collect more in a shorter space of time. If it switches to other flower types, it would have to learn where the pollen and nectar are on all of them, and apparently this is harder for the bees to do. Bees also get a visual "search image" of one type of flower which enables them to locate that flower type more easily. Bees conserve energy when collecting food by continually returning to the same *patch* of flowers and not spending too much time en route doing other exploring.

It's interesting to watch how a bee approaches a group of blossoms on the same plant. It seems to have two strat-

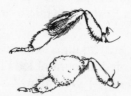

Detail of hind legs. Above, *pollen basket empty*; below, *pollen basket filled*

egies to avoid visiting the same one twice. On long spikes of flowers, such as loosestrife, the bees tend to land near the bottom of the spikes and work their way up in a spiral fashion. Flowers are often arranged in a spiral fashion on the plant as well. On large, flat-topped flowers, such as Queen Anne's lace, or composite flowers, such as daisies, where there are many individual florets offering food in the center, the bee tends to land on the edge and move in a spiral toward the center of the flower. On other less regularly arranged clusters of flowers, watch for yourself and see how well the bee covers all of the flowers without repeating. Do you see any pattern?

Other things that you can observe while watching bees at flowers is where they stick their mouthparts into the flower and how long they remain there. This may indicate the location of the nectar and how much is available. You can also watch the bees collect pollen, gathering it with their front legs and passing it back to their hind legs, where it is packed into small openings, called pollen baskets, for carrying back to the hive. Scientists have checked the pollen loads of bees to see how much of it is mixed from different species of flowers, and have discovered that only about 3 percent of the pollen in an individual bee's load is from more than one type of flower. This fact more than any other points to how flower-constant honeybees are.

Winter Insects

Observing Insects in Winter

Except for a few that migrate, all of the insects of spring, summer, and fall are still around in winter. The question is, where? And the answer lies in knowing what to look for and in keeping a watchful eye during leisurely winter walks. Actually, in winter, you are more likely to see the evidence of insects — their feeding places and built shelters — than the insects themselves, which are mostly buried in the ground, mud, or leaf litter. Searching for this evidence of insects in winter is like a treasure hunt, one that can become extremely enjoyable for many winters to come.

One of the most fruitful places to search is along the shrubby borders of fields and woods. Look among the bare branches of the shrubs for any bit of material attached to them, whether it be a leaf or something unrecognizable. If you find something, examine it to see if it is the work of one of the visible winter insects. Among the things you will find are several types of cocoons. The largest, two to four inches long, will be those of silk moths; smaller ones an inch long and wrapped inside a leaf will likely belong to the tussock moth. Tiny groups of cocoons, each no more than a quarter inch long, probably belong to braconid wasps. Another cocoonlike structure hanging off these twigs will be the larval cases of bagworms; they range from a half inch to over two inches long. If you find a tiny bit of curled-up leaf attached to a willow or poplar, you may have found the winter home of the viceroy butterfly. And if you find little holes all lined up along a twig, you may have discovered the eggs of tree crickets. Another treasure on twigs is the egg case of a mantid, a one-inch sphere of layers of hardened tan foam.

If you go into the woods, look first for dead or dying trees and check under their bark for the tunnel patterns of bark beetles or the intriguing "nest" of the ribbed pine borer. If you are near pine trees, examine the needles for bunches that are tied together with silk, forming a tube, and eaten off at their tips. These are made by the pine-tube moth and are especially common on white pine. And if you are near hickory trees, look beneath them for twigs or branches that are neatly severed; these are the work of the twig pruner and twig girdler beetles. Beneath the trees and on the surface of the snow, look for one of the few insects that is active in winter, the snow flea, an insect so small that groups of them look just like pepper sprinkled on the snow.

Finally, two very common and conspicuous insects in winter are the cattail moth and the goldenrod gall fly. The former is found within the seedheads of cattails, and the latter is found on the winter stems of goldenrod. All of these insects are among the most common to be found in winter.

WINTER INSECT LOCATION GUIDE

IF YOU ARE NEAR LOOK FOR	FIELDS	PONDS & STREAMS	WOODS	FIELD EDGES	HOUSES	BARE GROUND
SNOW FLEA			■			
TREE CRICKETS	■			■		
MANTIDS	■			■		
TWIG GIRDLER AND TWIG PRUNER BEETLES			■			
RIBBED PINE BORER						
BARK BEETLES						
VICEROY BUTTERFLY	■			■		
SILK MOTH	■					
CATTAIL MOTH		■				
BAGWORM MOTH	■			■		
WHITE-MARKED TUSSOCK MOTH			■	■		
PINE-TUBE MOTH			■	■		
GOLDENROD GALL FLY	■					
BRACONID WASP				■		
SPIDER EGG CASES			■	■	■	

SNOW FLEA

Relationships

The snow flea is a species (*Hypogastrura nivicola*) of insect in the family Hypogastruridae, which in turn is in the order Collembola, or springtails. Springtails are tiny insects, usually less than a tenth of an inch long. They are perhaps the most primitive of all insects and, along with the bristletails (order Thysanaura), are the only consistently wingless insects. Since they have no wings as adults, they look the same throughout their development except that they get bigger. Because of this they are often described as having no metamorphosis. They feed on decaying plants or animals and are often seen in large numbers on the surface of water or snow.

Life Cycle

Snow fleas overwinter as adults and are frequently active on the surface of the snow or on top of leaves. They mate

Snow fleas are the dots on the leaf. Life size

Habitat of snow fleas

in late winter and spring and the females lay eggs just beneath the soil surface. The eggs hatch and the nymphs feed throughout summer. It is not known whether there is more than one brood, but mature adults exist at least by winter.

Highlights of the Life Cycle

Snow fleas are hard to spot because they are so small. Winter is one of the best times to find them because they have matured and are highlighted on the surface of the snow. Even so, without the aid of a magnifying glass, you will only see small flecks of black appearing and disappearing on the snow as the insects hop about. Fortunately, the insects are usually in large groups, making them more conspicuous, and you can watch their movement and study what places they seem to favor.

How to Find Snow Fleas

Pick one of the warmer days of winter, when the temperature is above freezing, and keep your eyes open for patches of snow in the woods that seem slightly dirty, as if pepper were sprinkled on top. Look closely at these areas, perhaps putting your hand in among the flecks. If some of the flecks seem to disappear before your eyes or appear on your hand, you have found a group of snow fleas. The insects are so small and jump so quickly that they seem to just disappear when they move.

What You Can Observe

GROUP BEHAVIOR

Just as you may be surprised to find that the most common mammal on land is the tiny shrew, you may be equally surprised to find that the most common insect on land is the springtail. Even if you have an average interest in nature, don't be ashamed if you haven't seen or even heard of either shrews or springtails. Most people haven't.

Several species of springtails, among which the snow flea is the most common, are known to gather into large groups from late fall until early spring. For the most part, these

Magnified version of a snow flea

groups remain under leaf litter, feeding on decaying vegetable matter. But at other times they can be observed on
the surface of the leaf litter or snow, making long (for them)
overland migrations of up to twenty-five meters over a
period of a couple of days. These migrations are made in
groups of one-half to one million insects. The groups stay
fairly concentrated and in a roughly rounded mass as they
move. The surface ones hop, while the vast majority, just
beneath the leaf litter, may crawl. At night they stop moving and go under the leaf litter. Early the next morning
they start again. After a few days these groups can no longer
be found. Possibly they disperse, although it is not known
for sure.

Snow fleas often spend the warmer winter days on the
snow surface. They emerge from under the snow near tree
trunks or weed stems where the snow has melted away, and
then return through the same places at night. Snow fleas
can also be seen on icicles or on pond surfaces as the ice
melts. There are various colors of Springtails: the snow flea
is gray, but other species are orange, golden, white, brown,
green, blue, or red.

INDIVIDUAL BEHAVIOR

Individual snow fleas are only a few millimeters long, so
you have to get to within a foot of them to even distinguish
individuals. Put your hand down close to them and soon
you will see them suddenly on your hand, small dots that
continually appear and disappear. If you lift your hand up
six inches or more from the snow surface, they will all
soon vanish from your hand.

This curious behavior is the result of one of the snow
flea's modes of locomotion. On a latter segment of its
abdomen are two prongs that bend around and underneath
the insect. They are held in place there by two small hooks

on the belly of the insect. When the hooks open, the prongs spring out, push against the ground, and propel the insect through the air. This leaping behavior gives this insect its common names: springtails and snow fleas (since fleas jump).

Another interesting physiological feature of these insects is a small tube that protrudes from a front segment of their abdomens. Though its total function has yet to be determined, it seems to help the insect hold on to structures and also to enable it to absorb moisture.

Both the tube and the springing mechanism are much too small to see with the naked eye.

The presence of this tube, the springing mechanism, and the fact that these animals have only six abdominal segments, rather than the usual eleven as in most insects, has led some scientists to question whether springtails should be classified as insects.

TREE CRICKETS

Relationships

Tree crickets are a group of insects in the family Gryllidae, or crickets, which in turn is in the order Orthoptera, or crickets, grasshoppers, roaches, and mantids. Tree crickets differ from field crickets in having a slender body and being pale green rather than brown or black. They tend to live among woody plants and often feed on other small insects, such as aphids. Their chirping is particularly beautiful, being rich and full.

Habitat of tree crickets

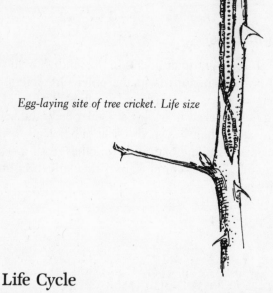

Egg-laying site of tree cricket. Life size

Life Cycle

Tree crickets overwinter as eggs that have been inserted into woody stems. In spring the eggs hatch, and the nymphs start to eat small aphids. They shed their skins several times as they develop. In mid- to late summer, they are mature and have wings. Males call to attract females. After mating, the fertilized females lay eggs in the stems of woody plants, where they overwinter.

Highlights of the Life Cycle

These crickets are not easy to see, for they are well camouflaged and spend a great deal of time up in trees. But they can certainly be heard in late summer and fall, since the males are loud singers. In winter there is also a good chance of finding their egg-laying sites. These are fascinating and very likely the most commonly seen stage of this insect's life.

How to Find Tree Cricket Eggs

Look among shrubs along the edges of fields or roads. Check especially the upper stems of raspberry, elderberry, or swamp dogwood for a slit in the bark two to four inches long, with tiny holes, each the size of a pinhead, all down the center of it. The whole egg-laying site looks a little like a zipper in the bark. The twigs used are usually about the size of a pencil or slightly larger.

What You Can Observe

THE EGG-LAYING SITES

Once you have found an egg-laying site, determine whether it is fresh and contains eggs, or is old and has been abandoned. Fresh sites that contain eggs will be on new, vigorous twigs, and the holes in the site will be filled. Old sites are on older twigs that have often died at or above the egg site. The egg holes will be empty and the exposed wood around them will be weathered and gray. Dead or broken twigs on the food plants are often a good sign of old egg sites.

To get a better look at the egg site, split it open with a penknife. You will see sets of parallel tunnels placed at right angles to the branch axis. At the base of the tunnel you may see some small deposit. This is a small bit of feces that the female, for some unknown reason, deposits in the hole before placing the egg in. After this she lays the egg

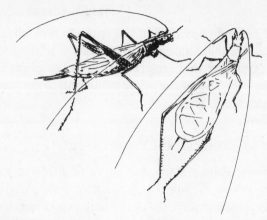

Left to right: *Black-horned tree cricket and snowy tree cricket*

in the hole and covers it over with a substance that seals it off from the weather. If you happen to open a fresh egg site, you may find the yellowish eggs inside. In this case, try to leave the eggs alone so that they have a better chance of hatching in spring.

Although adult tree crickets are beneficial to fruit growers because they eat aphids, the egg sites can be damaging because they kill fruit-bearing twigs or allow entry of various fungi, which in turn kill the plants.

TYPES OF TREE CRICKETS

The most famous tree cricket is the snowy tree cricket, because of its steady singing in choruses in late summer and fall and the predictable number of times it chirps in relation to the temperature. It has a low and very steady call, and if you count the number of chirps it gives in fifteen seconds and add that to thirty-seven, you have a close approximation to the temperature in degrees Fahrenheit. Temperature affects the calling rate of all crickets. The snowy tree cricket just happens to vary its rate in proportion to the Fahrenheit scale.

This species lays its eggs in the stems of fruit trees, but usually places them singly and irregularly over the twigs and branches, unlike the black-horned tree cricket, not as well known for its singing, which is the main species that lays its eggs in the long zipperlike fashion. The black-horned tree cricket got its name from its long, dark colored antennae. The rest of it, like the snowy tree cricket, is pale green.

MANTIDS

Relationships

Mantids are a family (Mantidae) of insects in the order Orthoptera, or crickets, grasshoppers, roaches, and mantids. All insects in this order are similar in that they have a first pair of wings that are straight and narrow and a second pair which fold and open like a fan. Mantids are large insects, two to four inches long, with specially adapted front legs that can reach out and capture other insects. They all make large, hardened foam casings around their eggs.

Life Cycle

Mantids overwinter as eggs placed in a mass on twigs, plant stems, or even buildings. The young nymphs hatch in late

Habitat of praying mantis

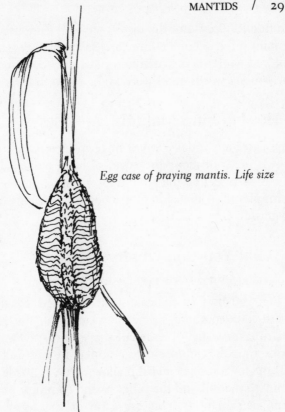

Egg case of praying mantis. Life size

spring or early summer, disperse, and feed on insects they catch in their front legs. They are full grown and winged by late summer. They mate, and the fertilized females lay their eggs in groups of several hundred, surrounded by a foam that hardens. The insects overwinter in this stage.

Highlights of the Life Cycle

As mantids get larger in late summer, there is a good chance you may see one perched among vegetation and waiting for prey. But the insects are well camouflaged and remain still most of the time, and so are hard to see. When fall and winter come, mantid egg cases become conspicuous among weeds and shrubs and are fun to find.

How to Find Mantid Egg Cases

Egg cases are usually found in open areas, such as overgrown fields or along the edges of fields and roads. Look one to four feet off the ground for masses of hardened, tan foam about an inch long, attached to the stems of weeds or grasses.

What You Can Observe

DIFFERENT TYPES OF EGG CASES

When you find an egg case, note its shape, for there are about four common species of mantids in the East and each makes a slightly different shape of case. Two species make roughly globular cases, similar in size and surface texture to a roasted marshmallow. One is made by the Chinese mantid and the other by the common praying or European mantid. The egg case of the former is all rounded, while that of the latter is clearly flattened on one side. The egg case of the narrow-winged mantid is shaped like a cannister — more cylindrical than round. All three of these mantids may be seen in the North as well as in the South. Another common mantid is the Carolina mantid, and it appears most frequently in the South. Its egg case is a flattened oval mass that is spread out along its support.

EGG-CASE CONSTRUCTION

If you opened an egg case you would see that it consists of three separate parts. Underneath, and nearest the support, are rows of five to ten eggs lined up across the axis of the case. On both sides of the eggs and forming the bulk of the case is a tough, frothy material uniformly laid down. Right down the center of the case is another distinct covering that is a series of overlapping scales. This has been called the "zone of issue," for this material is constructed to allow the young mantids an easy way to break out of the case (the sides are impenetrable).

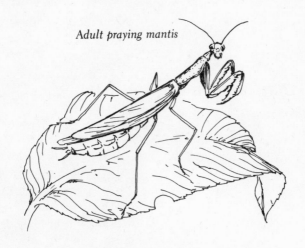

Adult praying mantis

The female mantid lays down all of this marvelous structure continuously, rather than in stages, which says a lot for the complexity of her egg-laying organs. The material around the eggs has air frothed into it by whipping actions of appendages on her abdomen. The material is soft and sticky at first but soon dries to become hard, stiff, and water-repellent. The eggs are laid in late summer, and a female may make up to fifteen or more egg cases.

The egg case seems like a perfect protection for the egg, but, as usual, there is a small ichneumon wasp, *Podagrion mantis*, which has a long ovipositor that is able to break through the case and deposit eggs in the eggs of the mantid. The wasps mature in the egg case and hatch the next summer. If you bring an egg case inside in spring to watch the young mantids emerge, you may see many of these parasites emerge as well.

TWIG GIRDLER AND TWIG PRUNER BEETLES

Relationships

The twig girdler (*Oncideres cingulata*) and the twig pruner (*Hypermallus villosus*) are species of insects in the family Cerambycidae, or longhorned beetles, which in turn is in the order Coleoptera, or beetles. Longhorned beetles are, for the most part, long beetles (an inch or more in length) with antennae longer than half the length of their bodies. Many have longitudinal or horizontal bands of color across their wing covers. The larvae of most burrow into woody plants. These two species have the unusual habit of cutting off the twigs in which the larvae feed.

The work of twig pruners: a.–c. are twig ends, smooth or spiraled, cut by larvae; d. shows area where larvae ate wood. Life size

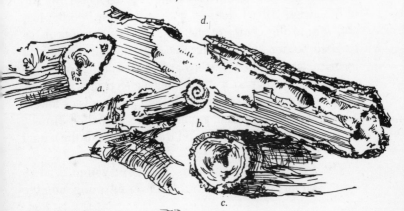

Habitat of twig pruners and girdlers

Life Cycle

The life cycles of these species differ. The twig pruner overwinters as a pupa within the largest end of a fallen twig. In spring it emerges as an adult. The adults mate in summer, and the fertilized female lays eggs near the tips of live twigs. After hatching, a larva burrows into the twig and toward the center of the tree. By late summer it has reached portions of the twigs that are one-half to one inch in diameter, and here it stops and starts cutting in a spiral fashion perpendicular to the branch, effectively cutting off the branch except for the bark. It then burrows back toward the tip of the twig and pupates. The branch is usually broken off in the fall winds and the pupa spends the winter in the fallen branch.

The twig girdler overwinters as a larva within a fallen twig. The larva pupates within the twig in spring, and the adult emerges by summer. After mating, females lay their eggs on the tips of twigs and then move down the twig and cut a circle around it so that only a small bit of wood remains. They then leave. The larvae develop within the twig by eating the wood over a period of two to three years, during which the twig usually falls off the tree.

Highlights of the Life Cycle

You will probably never see a live example of one of these adult beetles, but you have a very good chance of finding their handiwork in the form of the pruned or girdled twigs. These are fascinating to look at for they are so neatly severed. You can also trace the life of the larva by cutting open the twig and seeing where it has eaten away the wood.

How to Find the Work of the Twig Pruner and Twig Girdler

Look under oaks, hickories, or apple trees for fallen twigs one-half to one inch in diameter. Their ends should show evidence of being neatly cut off and not just broken off or rotted. The twigs often have leaves still attached, since they were cut off in summer before the leaves were shed.

What You Can Observe

TWIG-CUTTING BEETLES

There are a number of beetles that have the intriguing habit of cutting twigs off trees. These twigs are usually neatly severed and thus can be easily distinguished from other twigs that have fallen naturally. The work of the beetles is always a treat to discover and is surprisingly easy to find if you know when and where to look for it.

The twigs can be cut either by the twig pruner (who cut in their larval stage), or by the twig girdler (who cut in their adult phase). Each makes a distinctive type of cut on the severed twig.

For both, these habits of twig-cutting are believed to create better conditions for certain stages of the insect's development. The twig pruner makes the twig fall to the ground before its overwintering as a pupa, which would certainly protect it from woodpecker attack as well as from the extreme cold. The adult twig girdler kills the twig in which the eggs develop, possibly creating conditions that make the wood more palatable to the larvae. These explanations are just guesses. The real purpose may be even more interesting.

EVIDENCE OF TWIG PRUNERS

These twigs or branches look as if they have been pruned with shears. The best way to find and distinguish them

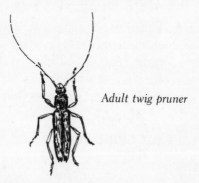

Adult twig pruner

from other fallen branches is to go to an area where there are several hickories or oaks. The first good clue is to look for fallen branches with leaves still attached, for the branches are usually pruned before the leaves fall. The branches range in diameter from one-half to one inch.

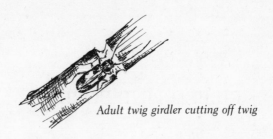
Adult twig girdler cutting off twig

Even if the branch does not have leaves, check the broken end, and if it is nearly cut off with a spiral groove, then it is one of the twig pruner's twigs. Sometimes the cut is so neat that it is just a smooth concave surface. Check out also large fallen limbs, for there are often smaller branches on them where twigs have been pruned.

By peeling away the bark just behind the pruned end, you should be able to see the tunnel of the larva, since it usually bores just beneath the bark. After the larva makes the pruning cut, it bores back into the outer portion of the twig and pupates in an enlarged chamber. If it is an old twig and the adult beetle has left, you should see a small exit hole coming from this chamber out through the twig bark. On particularly lucky days I have found as many as twenty-five pruned twigs in the space of a half hour under a grove of hickories.

EVIDENCE OF TWIG GIRDLERS

These branches have been cut starting from the outside and therefore generally have a ragged interior core where the branch broke off. The female, after laying eggs at the tip of the branch, crawls down and cuts the groove, then she makes transverse slits in the bark on each side of the groove. The groove and the slits cause the twig to die, and in some cases the twigs fall, especially on honey locust and persimmon. On other favorite trees of the female twig girdler — hickory, pecan and basswood — the twigs have more of a tendency to remain on the tree.

RIBBED PINE BORER

Relationships

The ribbed pine borer is a species (*Stenocorus inquisitor*) of insect that belongs to the family Cerambycidae, or long-horned beetles, which in turn is in the order Coleoptera, or beetles. Longhorned beetles are long and thin and their

Habitat of ribbed pine borer

Winter "nest" of ribbed pine borer. Life size

larvae feed within burrows in trees. The adults are often seen feeding on flowers in late summer. They can be roughly recognized by having antennae more than half the length of their body. Many are called borers, which refers to the larval habits.

Life Cycle

Ribbed pine borers overwinter as adults just beneath the bark of dead pine trees. In spring they emerge by chewing their way through the bark. After mating, fertilized females lay eggs on the bark of dead pines, and the larvae burrow through the bark and into the wood. By fall the larvae are ready to pupate and burrow to just underneath the bark. Each larva then makes a little circle of long wood shreds and pupates there. The adult emerges from pupation in fall but remains in the circle of shreds under the bark throughout the winter.

Highlights of the Life Cycle

As larvae these insects are buried within the wood, and as active adults they are flying about and well camouflaged, but in their winter homes, they can be easily found by pulling off loose bark. This also gives you a chance to see the marvelous winter "nest" they build around themselves.

How to Find Ribbed Pine Borers

Look for pine trees or cut pine logs with bark that is either loose or falling off, and gently pull it off. On the inside of the bark or on the exposed tree, look for an oval nestlike structure about one inch in diameter; a beetle may be nestled inside it. They are especially common on red pine.

What You Can Observe

HABITS OF WOOD-BORING BEETLES

Looking under the bark of dead trees always uncovers mysteries. Something that you will often find is the "frass" of wood-boring beetles. "Frass" is any material left behind as a result of feeding activity. In the case of wood-boring beetle larvae, this includes two types of material: digested wood that is excreted, and undigested wood that was just chewed off. The former is usually powdery and darker than surrounding wood, while the latter is like little wood chips or strands of wood and is the same color as surrounding wood.

Each species of wood-boring beetle has slightly different characteristics to its frass and different ways of depositing its frass in the tunnels that it bores. Some keep their tunnels clear by ejecting all frass; others plug entrance holes through the bark with the frass; some pack the tunnels behind them with it; and others, like the ribbed pine borer, even build structures out of it.

There are two families of beetles that are responsible for most of the large bored holes in our forest trees. They are the Cerambycidae, sometimes called round-headed borers or longhorned beetles, and the Buprestidae or flat-headed borers. The terms *round-headed* or *flat-headed* refer to the larva, which in the case of the Buprestidae has a large, flattened head, and in the case of the Cerambycidae has a round-shaped head, more normal for a larva. This difference makes the larvae of these two families easy to distinguish. There is no easy way to distinguish between the work of these two families of beetles except that the adult exit holes through the bark are usually oval in the Buprestidae, and round in the Cerambycidae. For the life habits of other Cerambycids, see the twig pruner beetles (winter) and the longhorned and soldier beetles (fall).

HABITS OF THE RIBBED PINE BORER

As a larva, the ribbed pine borer feeds in the wood of dead pines and leaves behind lots of long shreds of wood averaging about a quarter of an inch in length. When it is full

Adult ribbed pine borer

grown and about to pupate, it tunnels almost to the surface of the bark, then backs down to where the bark meets the sapwood and hollows out a flattened chamber. It then pulls off shreds of wood, each over a half inch long (twice as long as normal shreds) and arranges them into an oval shape that ends up looking like the top of a bird's nest. In this space it pupates, emerges as an adult, and then remains there, as an adult, through the whole winter. In spring it breaks through the last remaining bits of bark at the end of its exit tunnel and flies off to new trees. After mating in early spring, the female seeks out trees that have recently been cut, or have recently died, and lays her eggs in the bark crevices. The larvae feed under the bark, making large, irregular galleries and filling them with frass that contains both digested and undigested wood.

If you look for the beetles on trees that have been dead a long time, you will find mostly empty pupal "nests" with exit holes continuing through the bark. But if the tree is fresher, you may find the adult beetle in the pupal chamber. It is about three-quarters of an inch long, and speckled gray and white. It has very clear ridges on its fore wings. Whenever you look under bark, and you find animals there, be sure to try and replace the bark carefully just the way it was. Otherwise, most of the living things underneath it will die from exposure to the cold or the drying of the wind.

BARK BEETLES

Relationships

Bark beetles are a family (Scolytidae) of insects in the order Coleoptera, or beetles. Most are minute insects about an eighth of an inch long. All of them feed on the wood of trees and many of them make lovely patterns of tunnels as they feed in the layer of wood between the bark and the trunk; these species are often called engraver beetles. Other species leave conspicuous minute holes in the bark as they leave or enter, and they are called shot-hole borers. Still

Habitat of bark beetles

Bored galleries of bark beetles. Life size

others raise fungus in their tunnels and are called ambrosia beetles.

Life Cycle

Most bark beetles overwinter as mature larvae at the end of tunnels they have bored in trees. They pupate in spring and the adults emerge in early summer. They bore out through the bark and fly to new trees. Either the male or female, depending on the species, bores into the new tree and makes a chamber. The opposite sex soon enters the hole and the two mate in the chamber. Following this, the female makes a channel in the wood with niches on each side in which she lays eggs. The eggs soon hatch and the larvae start to eat away the wood, either in separate tunnels or together in a single cavity. When mature, they make a slightly enlarged space in the wood and pupate. After changing to adults, they bore their way out through the bark. There may be one or two broods, the larvae of the last brood overwintering in their tunnels.

Highlights of the Life Cycle

Most of the bark beetle's life is spent inside wood, and only in early summer do the adults emerge and fly to new trees. If you are near coniferous woods at this time of year, you may see the beetles land on you, but they are minute and rectangular, and unless you were very aware, you would probably dismiss them as pieces of debris. The beetles' life habits are recorded under the bark of trees where they have been living. Therefore, in winter, the whole life cycle is a highlight, and even though you may not spot the actual insect, you are bound to see evidence of all its life stages.

How to Find the Work of Bark Beetles

Look in older forests where there are dead trees and fallen branches. Pull loose bark away and look on the exposed wood of trunks or branches for carved patterns in the bark or wood. Some of the patterns are very regular. These are probably those of bark beetles. The channels are very narrow, usually no more than an eighth of an inch wide. Bark beetles make tunnels in both deciduous and coniferous trees, but their work is more common on conifers.

What You Can Observe

THE TUNNELS OF ENGRAVER BEETLES

One especially enjoyable winter activity is walking through woods and looking for the various tunnel patterns of bark beetles on dead or dying trees. With no foliage to obscure your vision, these trees are easier to spot and distinguish from healthy, living trees. The galleries appear as little grooves all over the surface of the wood under the bark, or

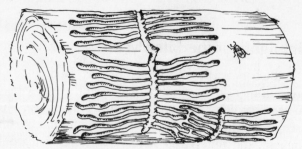

Exposed galleries and bark beetle on log

also in the undersurface of the bark, some forming marvelous regular patterns.

If you look closely at bark beetle tunnels, you can see how they reflect the behavior and life histories of their makers. You will see basically two types of tunnels: those of uniform width, which are made by adult beetles (see Life Cycle), and those that increase in size, which are made by the larvae as they grow.

The uniform tunnel was made by the adult female beetle after she had mated with the male in a slightly enlarged chamber at one end of the tunnel. Usually you can see little niches on each side of the egg tunnel. These are the egg niches and they were also made by the female to provide places for her to lay her eggs in. After the female makes the egg tunnel and egg niches, and lays the eggs, she exits directly through the bark.

The tunnels made by the larvae come off the egg niches, and are small and shallow and radiate to each side of the main egg tunnel. The reason for their distinguishing characteristic of getting wider as they move away from the egg niches is that the larvae are growing as they are making them. At the end of these larval tunnels is usually a little hollowed-out section where the larva pupated before chewing out through the bark and emerging as an adult.

Bark beetles make their tunnels right at the point where the outer layer of wood and the inner layer of bark meet.

For some species, their excavating touches both bark and wood; for others, just the wood or just the bark. Therefore, when you examine loose bark, you should look at the wood underneath, for it may reveal other features of the tunneling.

In some species, a single female forms a single egg channel. In others, two or more females mate with a single male and form two or more egg channels coming from a central spot.

As you discover more and more examples of bark beetle work, you will begin to see patterns. Certain types of tunnel patterns will always be found on certain trees. This is because the beetles in many cases always choose a specific host tree, and sometimes even a particular size branch on that tree.

THE TUNNELS OF AMBROSIA BEETLES

Ambrosia beetles are a type of bark beetle, but their tunnels go into the center of the wood rather than remaining on the surface. Because of this, there are very few ways to find the tunnels. The best way is to look at split logs, especially of apple trees, and search for long, even-width tunnels that are stained black. The tunnels are the same size as the egg tunnels of other bark beetles.

Exposed tunnels of ambrosia beetles: small, vertical "cradles" on left; longer, horizontal adult tunnels on right

The tunnels are made by the adult males or females, depending on the species. They are long and branching, and at their tips deep inside the tree, they have a cluster of shorter tunnels, each about a quarter inch long, coming off them at right angles. These have been called cradles, for the female lays an egg in each one, and the young complete all of their development in them before leaving. There are no larval tunnels, because the young, instead of eating wood on their own, are fed bits of fungus by the adults. This fungus is grown by the adults in the longer tunnels, which is why they are stained black. Fungi are pushed into the cradles until the entrance is plugged. As soon as the larva eats through it, more is plugged in. The larva does not leave the tunnel until it has pupated and is an adult. Most species of ambrosia beetle raise only one species of fungus, and when the adults move to new trees, they are believed to carry spores of this fungus with them.

VICEROY BUTTERFLY

Relationships

The viceroy butterfly is a species (*Limenitis archippus*) of insect in the family Nymphalidae, or four-footed butterflies, which in turn is in the order Lepidoptera, or moths and butterflies. The Nymphalidae are mostly large butterflies, and all have very small front legs that are held up and out of sight against the body. The genus *Limenitis* contains several common butterflies, each with a wingspan of about two and a half to three inches. They all have the fascinating habit of overwintering as larvae in rolled leaves fastened to their food plant by silk.

Habitat of viceroy butterfly

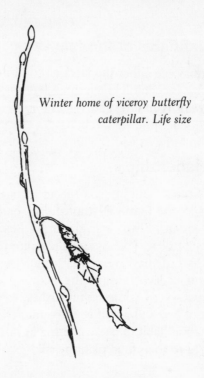

Winter home of viceroy butterfly caterpillar. Life size

Life Cycle

The viceroy butterfly overwinters as a partially mature larva rolled up in a leaf attached to the food plant. In spring the larva emerges from the leaf and feeds on new leaves. Its main food plants are willows and poplars. It matures in a few weeks and then pupates attached to a branch or leaf. Pupation lasts seven to ten days. The adults emerge, feed, and mate, and then the fertilized females lay eggs on the tips of leaves. The eggs hatch in four to eight days, and the larvae feed on the leaves. They mature in three to four weeks and then pupate. Adults emerge, mate, and start another brood, but the caterpillars from this brood over-winter. There are sometimes more than two broods, but the larvae of the last brood always overwinters.

Highlights of the Life Cycle

You may see either the larva or adult in spring or summer and this is certainly interesting, but by far the most exciting discovery is finding the viceroy's overwintering site. This is partly because we rarely have a chance to see any aspect of a butterfly in winter, but also because the butterfly larva seems to take so much care in its preparations for winter.

How to Find the Viceroy Larva in Winter

Look on the twigs of willow, poplar, or aspen saplings for any remains of leaves. If you find a leaf remain, one-half to one inch long, curled over into a tube, and secured to the twig with silk, then you have the winter home of the viceroy. They are not easy to find and are uncommon.

What You Can Observe

THE WINTER HOME AND ITS CONSTRUCTION

The winter home of the viceroy butterfly is one of the more difficult finds, but at the same time one of the most exciting. In order to locate it you have to be at least as good a botanist as the female butterfly, for she lays her eggs singly on the tips of only willow or poplar leaves. Look especially on willow shrubs, which are commonly found bordering meadow streams, or on sapling poplars, such as aspens, which can be found along highways or in abandoned fields. What you are looking for is any tiny bit of leaf still attached to the twigs. It will be about a half inch long and rolled into a tiny tube about an eighth of an inch in diameter. The best clue as to whether some leaf fragment is a viceroy winter home is to look at the point of attachment to the branch, where there should be a distinct band of white silk

From left to right: *larva, adult, and pupa of viceroy butterfly*

wrapped around the twig and the leaf stem. Inside this little cylinder of leaf is a tiny caterpillar, the overwintering stage of the butterfly.

The egg for this caterpillar is laid in late summer. After hatching, the larva eats at the tip of a leaf along each side of the midrib. After a while, this results in a leaf with its base uneaten and its tip just a bare mid-vein. The caterpillar, still less than a half inch long, curls the base of the leaf into a tiny cylinder and lines it with silk. It also wraps the leaf petiole with silk and connects it with a band around the twig so that when the other leaves fall to the ground, this one remains. It then crawls headfirst into its winter home. In spring, when the catkins bloom and the leaves emerge, the young caterpillar will come out and start to feed on either of these new growths.

Think twice before bringing the larvae of these butterflies inside in winter, for the chances are that the sudden change in temperature and the dry air in your house will kill them. It is best to leave them outside so that we have a chance to see the beautiful mature butterflies in the spring.

There are two other species of butterflies in this genus with the same overwintering habits as the viceroy, and you may discover their winter homes as well, although generally they are more difficult to find. The white admiral is a dark butterfly with a wide white band running down each wing. It lives in northern North America, and its winter homes are primarily on young birch trees. The red-spotted purple is also dark with blue dots on the upper edge of its wings and red spots underneath. It lives in the southern half of North America and its winter homes can be found on members of the rose family such as cherry, apple, and rose.

THE VICEROY AND MONARCH

The adult viceroy butterfly is easily confused with the monarch, which, although it is no relation, looks almost exactly like it. This is one of the most famous examples of protective mimicry. As the monarch larva acquires certain chem-

From left to right: *white admiral and red-spotted purple*

icals from eating milkweed, its black and red colors warn predators of its distastefulness. The viceroy is believed to be perfectly tasty and thus gets some advantage from looking like the untasty monarch. It has recently been discovered that some birds can eat monarchs without being affected by the poisons. How long they have been able to do this is not known. If it becomes more widespread, the viceroy will no longer gain any advantage by its mimicry.

There are several good ways to distinguish the two butterflies. The best way is to look at the hind wings. Both species have orange wings with black veins and a black rim, but the viceroy also has another black line crossing the veins at right angles. (See illustration at Monarch Butterfly, fall section.)

SILK MOTHS

Relationships

Silk moths are a family (Saturniidae) of insects in the order Lepidoptera, or moths and butterflies. Their common name comes from their copious use of silk in their cocoons. They are our largest moths, many with wingspans of up to five inches; even so, the adult moths are rarely seen, for they are only active at night and are not usually attracted to lights, where we might see them.

Habitat of silk moths

Promethea cocoon on buttonbush. Life size

Life Cycle

Silk moths overwinter as pupae in cocoons. In late spring the adults emerge and mate, and the fertilized females lay eggs in small clusters on the leaves of their food plants. Most silk moths are generalists and feed on several species of plants; many prefer cherry. The eggs hatch in ten to twelve days. The young often start by feeding together and then, after several weeks, feed on their own. By late summer, the larvae are mature and up to four inches long. They build a cocoon and overwinter inside it.

Highlights of the Life Cycle

Just as the adult moths are rarely seen, the larvae are also not commonly found. Pupation is probably the most conspicuous stage of this insect's life, for the cocoons are about two to four inches long, and those of several common species are attached to the twigs of trees and shrubs. When the woody plants lose their leaves, the cocoons become easier to see.

How to Find Silk Moth Cocoons

One of the great treasure hunts of winter involves looking for the cocoons of our common silk moths. Although these cocoons are the largest and most conspicuous of any moth, they are fairly uncommon and hard to find, so it can be frustrating if you go out on a walk specifically to find one. It may also be that populations of these moths have decreased; many naturalists believe the cocoons were much more common in the early 1900s. They are most often found attached to twigs on trunks, or shrubs, or the lower branches of trees. Check any object three to four inches long attached to a twig, and if you find it is composed of silk, you most likely have a cocoon. When the cocoon is jostled, you may be able to hear or feel the pupa rattle inside.

What You Can Observe

TYPES OF COCOONS

The cocoons of silk moths are tough and papery and woven of silk, and often incorporate leaves. The most easily spotted type of cocoon is the hanging type made by the promethea

Female promethea moth and larva

and cynthia moths. When these caterpillars are ready to pupate, they choose a particular leaf on their food plant, cover the leaf petiole with silk, and then firmly attach it to the twig. After this, they form a cocoon out of the leaf, curling the leaf lengthwise around their body. Lots of silk is used and often the whole leaf is also covered. The cocoon is about two inches long, one-half to three-quarters of an inch in diameter, and tapers slightly at each end. Of the two species that make this type of cocoon, the cynthia moth is less widely distributed because it was imported from China and feeds only on the ailanthus tree, which was also imported from China and grows mostly in the waste areas of large cities. So if you find one of these cocoons on the branches of ailanthus, you can be sure that it is a cynthia moth. If you find one of these cocoons on any other shrub or tree, especially a cherry, spicebush, buttonbush, sassafras, tulip-tree, or sweetgum, then it is a promethea moth. This is our native species and its cocoon is so similar to that of the cynthia that it is usually only possible to distinguish the two species by their food plant or by raising them to the adult stage.

A second type of cocoon found in winter is broadly attached to twigs rather than hung from them, and although it sometimes incorporates leaves, is not so carefully wrapped in them. This is generally a larger cocoon than the hanging type but harder to see since it blends in with the branches of the plants. It is about three inches long and about an inch in diameter at its middle. This is the cocoon of the cecropia moth. Its larva feeds on a variety of common plants, including cherry, maple, box-elder, apple, elderberry, birch, and willow.

A third cocoon you may find is about an inch long and three-quarters of an inch in diameter, and contains leaves. Only occasionally attached to branches, it is usually made on the ground. This is the cocoon of the polyphemus

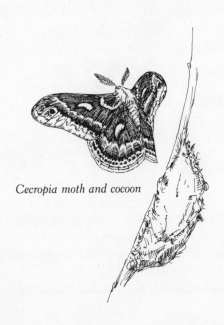

Cecropia moth and cocoon

moth, and the food plants it is found on or near include oak, maple, birch, and hickory.

In each of these three species, the adult moths emerge in spring by secreting a fluid that dissolves the silk of the cocoon. This is extremely important because, as you will

Polyphemus moth cocoon, larva, and adult

see if you ever have occasion to tug at an old cocoon, its outer covering is incredibly tough.

Many people collect these cocoons in winter as trophies or with the hope of watching the adult moths emerge from them in spring. Although this can be fun and educational,

I do not recommend it, for no one really knows how to recreate the winter and spring conditions that are correct for the moth, and little is known about how collecting cocoons has affected the population of these moths.

PARASITES OF SILK MOTHS

Sometimes you will come across a silk moth cocoon that, when lightly shaken, rattles as if there were a whole lot of small capsules inside. At other times you can see that the cocoon is only a thin layer of silk and that the caterpillar never completely transformed to a pupa. The reason for both of these occurrences is that the moth larva has been parasitized by an ichneumon wasp in the genus *Agrothereutes*. The wasp is attracted to the caterpillar when it is spinning its cocoon, possibly smelling the fresh silk. It lays its eggs in the caterpillar and the wasp larvae complete their development there. The caterpillar may or may not finish its cocoon before it is killed. About one-third to one-half of the cocoons I find have been parasitized.

CATTAIL MOTH

Relationships

The cattail moth is a species (*Lymnaecia phragmitella*) of insect in the family Tortricidae, or leaf rollers, which in turn is in the order Lepidoptera, or moths and butterflies. The leaf rollers are known and named for their larval habits, which often include rolling up or tying leaves together with silk as they feed on them. The cattail moth modifies this habit somewhat by tying together the dispersal filaments of cattails during fall and winter. The adult moths of this family are small and brown and difficult to distinguish from the many other moths that fly about in the evening.

Habitat of cattail moth

Cattail seedhead that looks normal but is actually tied together with silk from cattail moth larva. Reduced to one-half life size

Life Cycle

Cattail moths overwinter as larvae inside the seedheads of cattails. In spring they resume their eating of the cattail seeds and when mature make a cocoon in the seedhead and pupate. Pupation lasts about a month. The adults emerge and mate, and the fertilized females lay their eggs on the developing cattail flowerheads. In a week or two the eggs hatch and the young feed on the developing seeds. In late summer and fall, when the young are still only partially developed, they tie together the dispersing seeds of the cattail with silk and then remain among them through the winter.

Highlights of the Life Cycle

The easiest time to find these insects is in winter when they are inside the fluffy seedheads of cattails. At this time you can find the larva and see the way it has tied the seeds together. You may also be able to examine the same seed-head for a cocoon in spring.

How to Find the Cattail Moth Larva

Go to a marshy area where there are cattails and look for seedheads that have a great deal of fluff still attached to them. Examine these by gently pulling them apart. If the fluff is being held together with silk, then you have the winter home of the cattail moth larvae. They are very common and easy to find. The larva can be found inside the fluff.

What You Can Observe

THE LARVAL WINTER HOME

This is a common little moth with worldwide distribution that you have probably never heard of. But I am sure that once you learn about its life habits you will never forget it. It can be found in almost any patch of cattails, the tall marsh plants that have at the tips of their stalks a soft brown seedhead about the size and shape of a large cigar. These attractive seedheads normally begin to break apart in early winter and disperse their seeds in the wind to new areas. By late winter, some of the seedheads are suspiciously different. They look much like the normal seedheads that are dispersing seeds, except that when you shake them or blow on them, no seeds come off. By early spring they are even more conspicuous, for the fluff and seeds still remain attached to the stalks as in fall, while on normal cattails only the stalks remain.

The reason for all of this is the larvae of the cattail moth. The adult moths lay eggs on the maturing seedheads of the cattails in July and August. The larvae soon hatch and begin feeding on the inner portion of the flowers. As the seedhead dries, the larvae burrow farther in and feed on the matured seeds. Meanwhile, they lay down lots of silk trails all over the seedhead, which keep the down and seeds from dispersing. This benefits the moth in two ways: it keeps its food from being blown away and it holds together the down, which then acts as insulation for the larvae through the winter.

The larva is about half grown as winter approaches, and it remains in the cattail seedhead throughout the cold months. At this time, if you open up one of these seed-heads, you are likely to find the larvae inside. There may be from one to about fifty in a single seedhead. They will be about a quarter inch long and yellowish with fine brown stripes down their backs. In spring they become active again and continue feeding on the seeds until about May or June, when they make a tough cocoon of white silk, either in the down or, more rarely, in the plant stem, and then pupate. In about a month the adults emerge, and just after emergence rest on the outside of the down seedhead. In July, at cattail marshes in the early evening, you may see the adult females landing on the flowerheads to lay their eggs. The adults are about half an inch long and yellow-brown, with somewhat pointed wings. The best clue to their identity is their size and their presence on cattails.

There is another species of moth, even smaller, with identical larval habits, called *Dicymolomia julianalis*. It is not worldwide in distribution but mostly limited to the southern United States. It differs from the cattail moth only in that it always bores into the stem to pupate.

BAGWORM MOTHS

Relationships

Bagworm moths are a family (Psychidae) of insects in the order Lepidoptera, or moths and butterflies. The family is named for the habits of the larvae, which construct little bags of plant debris and silk to live in during their whole larval life, just poking their heads and feet out to feed and move.

Habitat of bagworms

Bag of evergreen bagworm. Life size

Life Cycle

Bagworm moths overwinter as eggs within their parent's bag. After hatching in spring, the larva leaves this and starts to create its own case or bag around itself as it feeds. The bag is continually enlarged as the larva goes through its five developmental molts. When the larva is mature, it fastens the bag to a twig or other secure object and pupates inside the bag. Pupation lasts several weeks, after which males emerge and seek out females, who emerge from their pupae but not from their bags. The male fertilizes the female through the tip of the bag. She then lays eggs inside the bag, exits, and dies. The eggs overwinter inside the bag.

Highlights of the Life Cycle

Chances are that you will never see an adult bagworm moth, since the females remain in their bags and the males are on the wing for only a very short while. However, you can observe the bags in winter, when the leaves are gone from the trees, and the bags, which are attached to branches, are more conspicuous. This is a time when you can get a close look at the marvelous construction of the bags, and also tell if the bag belonged to a male or female moth. You may also discover eggs inside the bag.

How to Find Bagworm Moth Bags

The most conspicuous bags are those made by the evergreen bagworm. They are about one to two inches long and are attached at their tips to twigs. They have fine twigs tied to the outsides of the bags with silk. Look for them among the branches of red cedars or on the lower branches of leafless shrubs or trees. Sometimes the swaying of the bags in the wind will draw your attention to them.

What You Can Observe

THE BAGS

The bags of the bagworm moths are constructed by the larvae and contain silk and material from plants that the larvae have been feeding on. The most commonly found larger bags belong to the Abbott's bagworm (*Oiketicus ab-*

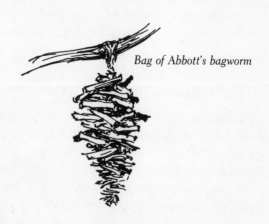

Bag of Abbott's bagworm

Bag of Eurycyttarus confederata

boti) and the evergreen bagworm (*Thyridopteryx ephemer-aeformis*). Both of their bags can be as large as two and half inches. The difference between them is that the twigs that cover the Abbott's bagworm bag are placed horizontally, while those on the evergreen bagworm are vertical. Both larvae are fairly general feeders and their bags can be found hanging on a variety of evergreen or deciduous trees or shrubs. A common smaller species is *Eurycyttarus confederata*. Its bag is usually about one-half inch long, covered with straight, short twigs, and is often found attached to the outsides of houses. The larva feeds on shrub and flower leaves.

All bagworm bags in winter will be attached to their support by silk. The attached end of the bag was the top where the head and feet of the larva used to protrude as the larva moved about and fed. At the bottom of the bag there will be a small opening. This is where the shed skins and feces of the developing larva were expelled.

If you open up the bags that you find, you will notice that some are empty while others contain yellowish eggs. The former bags belonged to male moths, and at the bottom of these bags you may also see an empty pupal skin halfway out the bottom hole. This is left by the male moth, which, while still in the pupa, wriggles down so that it is partially out of the bag and then emerges from the pupa.

The female bags contain eggs and also the remains of

Male evergreen bagworm adult

the pupal skin, because the female emerges from the pupal case but remains in her bag through mating and egg-laying. Only after that does she leave the bag.

Any bag that has something in it other than what is described above has more than likely been parasitized. Evidence of parasites would be several smaller cocoons inside the bag, or the bagworm larva eaten away. There is at least one ichneumon wasp, called *Ictoplectis conquisitor*, known to parasitize the larger bagworms.

ADULT BAGWORMS

Although you may never see an adult bagworm moth, they have some features that are unusual and worth discussing. The female is always wingless and in some cases also lacks mouthparts, legs, antennae, or eyes. So, rather than looking like a moth, she may look more like a large maggot. The male, on the other hand, has wings and feathery antennae. The latter may be to help him locate the female through scent. For some unknown reason, the scales on the wings of some species are only loosely attached and rub off as the male emerges, leaving its wings almost clear.

WHITE-MARKED TUSSOCK MOTH

Relationships

The white-marked tussock moth is a species (*Hemerocampa leucostigma*) of insect in the family Liparidae, or tussock moths, which in turn is in the order Lepidoptera, or moths

Habitat of tussock moth

Tussock moth cocoon with eggs on top. Life size

and butterflies. The common name for the family comes from the appearance of some of the larvae, which have several groups of especially long hairs, or tussocks, among their shorter ones. Two of the more famous members of the family are the gypsy moth and the white-marked tussock moth.

Life Cycle

The white-marked tussock moths overwinter as eggs placed on the abandoned cocoon of the female. In spring the eggs hatch and the larvae crawl about and feed on leaves from a wide variety of woody plants. In four to six weeks, they are mature and start to build cocoons in which to pupate. The cocoons are attached along their length to twigs. Pupation lasts about two weeks. The adult females are wingless and crawl out of their cocoons and remain on the outside. Males have wings and fly to females for mating. The females lay their eggs on the outsides of their cocoons. There may be one to three broods, and the eggs produced by adults of the last brood overwinter.

Highlights of the Life Cycle

You are not likely to see the adults, for they are nocturnal, but during the summer months you may notice the caterpillars because of their remarkable appearance: they have two long tufts of hairs at each end and four shorter tufts along their backs. When full grown, they are about one and a quarter inches long. Winter provides a good chance to see the cocoons, which, in some cases, have eggs laid right on top of them.

How to Find White-Marked Tussock Moth Cocoons

Go to the border of fields or roads where there are lots of shrubs and look among the twigs of the shrubs for any bit of dried leaf still attached to the branches. If you find one, then look inside it and see if it contains a cocoon covered with tan hairs. The cocoons are only about an inch long. There may also be silk fastening the leaf to the twig.

What You Can Observe

THE COCOONS

The search for cocoons among the winter trees and shrubs is a marvelous seasonal ritual. Those of the tussock moth are among the most common. They are found on bits of leaves that remain attached to the branches. This is because the caterpillar often secures the leaf to the twig with silk before making the cocoon. You will find two main types of tussock moth cocoons. Both are oblong and made of small tan hairs and silk, but one is about a half inch long while the other is over an inch long. The smaller ones belonged to the males. The larger ones were used by the

Cocoon, adult female moth above, male below

females, and you will find on them a hardened foamy material that contains eggs. To understand this unusual occurrence, we need to follow the events of the insect's life through the year.

In spring, small caterpillars crawl out from the eggs and begin to eat leaves and grow. After four to six weeks, they pupate in small cocoons made from silk combined with the hairs from their own bodies. In about two weeks, when the adults emerge, an amazing feature of these moths becomes apparent: the females have no wings. Also, the male's antennae are long and feathery, while the female's are short, single filaments.

These differences reflect the different life habits of the sexes. The female crawls out of her cocoon but remains on it. It is very likely that she gives off an odor that the male senses with his feathery antennae. He flies to her, they mate, and she lays her eggs right on the empty cocoon. After she has covered the eggs with a frothy mass also containing some hairs from her body, she dies.

There is one interesting problem that these wingless females pose. Theoretically, the caterpillars that hatch from the egg case will remain on the same plant. When the females among these in turn pupate, emerge as adults, and lay eggs, all of their young will also be on the same plant. How does this species get dispersed to new areas and not crowd itself out on one plant? If the females could fly, this would solve the problem, but they don't; therefore, this insect must be dispersed in other than the adult phase. We know that they are not dispersed as pupae, and it is unlikely that they are dispersed as eggs. This leaves the larval stage. As mature caterpillars, they are too heavy and slow to move very far, but it so happens that when they are very small they can spin silk into the air and be carried by the wind to new plants. An important feature of the larvae in this type of dispersal is that they are general feeders, for they obviously cannot control where they land.

THE CATERPILLAR

The white-marked tussock moth caterpillar is a common and spectacular sight. It may be seen in early or late sum-

White-marked tussock moth caterpillar

mer. Its head is green and its body is lined with black and white longitudinal strips. Short, cream-colored hairs cover most of the body, longer black hairs project in bunches from its head and tail, and four dense tufts of hairs project from the top of the segments just behind its head. The hairs on a few species of caterpillars can cause a rash, but in this species they are harmless, so the young insect can be picked up and examined without fear.

PINE-TUBE MOTH

Relationships

The pine-tube moth is a species (*Argyrotaenia pinatubana*) of insect in the family Tortricidae, or leaf-rollers, which in turn is in the order Lepidoptera, or moths and butterflies. The larvae of many species in this family have the habit of

Habitat of pine-tube moth

Two bunches of needles tied together for the winter homes of pine-tube moth larvae. Life size

tying leaves together with silk for protection. Most are very small moths that would go totally unnoticed except for their larval behavior.

Life Cycle

Pine-tube moths overwinter as pupae within bunches of pine needles. Adults emerge in late spring and mate, and the fertilized females lay eggs at the tips of new needles on pines. The eggs hatch, and the larvae crawl out on the needles and, with silk, tie some together into a tube. They then crawl inside the tube and feed on the needle tips. They may eat through several tubes before maturing and pupating within one for the winter.

Highlights of the Life Cycle

You can find the work of this insect in any season, but it is particularly fun to discover the pupa within the tubes of needles in winter. The adult and egg stages are rarely seen.

How to Find the Pine-Tube Moth Pupa

Go to any white pine and check the needles at the tips of the branches, looking for some that are tied together into a tube with silk and chewed off at their ends. Such a tube is the present or past home of a pine-tube moth. The tubes are extremely common; sometimes there are hundreds on a single tree. Inside some of the tubes will be pupae.

What You Can Observe

THE TUBES OF PINE NEEDLES

Evidence of the pine-tube moth is common throughout the range of the white pine. Sometimes two or three tubes will be seen in a single cluster of needles. The tubes are usually about half the length of normal needles because the larva has eaten off their ends. If you carefully open one of the tubes, you will see that the needles, usually about five to ten of them, are tied together with silk, and then the resulting tube is lined with silk. Many of the tubes will be empty, for each larva uses two or three tubes during its development.

THE LARVA AND PUPA

If you look at the tubes in fall or early winter, you may find a pine-tube moth still in the larval stage. When feeding on the needles, the larvae have the marvelous habit of

A single needle sticking loosely out of a tube and being eaten by the larvae hidden inside

biting the needle off at half its length and then pulling it back into the tube to eat. Sometimes you may even see a tube with a single needle sticking loosely out of the end. This is where a larva is feeding on a needle. The larvae are about a quarter inch long and orange.

The pupae are the overwintering stage of the moths and they are orange brown and also about a quarter inch long. They wiggle slightly when touched.

GOLDENROD GALL FLY

Relationships

The goldenrod gall fly is a species (*Eurosta solidaginis*) of insect in the family Tephritidae, or fruit flies, which in turn is in the order Diptera, or flies. Flies in this family are about a quarter inch long and often have dark patterns on their otherwise clear wings. They are not to be confused

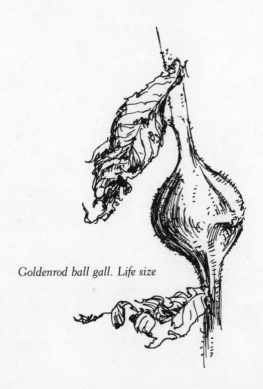

Goldenrod ball gall. Life size

Habitat of goldenrod gall fly

with the small flies that breed in any fruit that is left around; these are sometimes called fruit flies but are actually vinegar flies, or *Drosophila*. In the case of real fruit flies, the maggots bore into fruit and can cause a great deal of agricultural damage. A well-known pest and member of this family is the Mediterranean fruit fly.

Life Cycle

The goldenrod gall fly overwinters as a larva inside a round gall on the stem of goldenrod. In spring it pupates inside the gall and then emerges as an adult. After mating, the female locates new goldenrod stems that are just coming up in early summer and lays eggs on them. The eggs hatch and the larvae burrow into the stem and create a chamber in which they feed on the plant tissue. This causes the plant to grow a round deformation around the chamber of the insect, and this is called the gall. The larva overwinters in the gall.

Highlights of the Life Cycle

The most conspicuous product of this insect's life is the spherical gall that it causes to grow on goldenrod stems. Opening up a gall in winter will allow you to get a close look at the fly larvae and also see the structure of the gall.

How to Find the Goldenrod Gall Fly Larva

Go to a field where there was goldenrod in bloom in fall and look among the new dried weeds. Look for a round swelling about three-quarters of an inch in diameter on the goldenrod stems. This is the goldenrod ball gall and the winter home of the goldenrod gall fly. There may be more than one gall on a stem. The larva is inside the gall. This fly is common in only the northern half of North America.

What You Can Observe

THE GALL

Galls are an extremely common phenomenon. They are a deformation of plant growth caused by the physical actions or chemical secretions of insects. Usually the insects that cause the gall live in and feed on the plant material of the gall. By far, the majority of galls are made by minute flies and wasps whose larvae feed on the galls. Certain plants tend to have more galls. Each gall-forming insect chooses a particular species of plant, and its galls are distinctive in shape and color. Certain species of plants are particularly prone to attack by gall insects. Goldenrod is one of these. In most cases, galls do not substantially harm the plant.

The most easily seen of the galls on goldenrod is the ball gall made by this fly. It is a round swelling about three-quarters of an inch in diameter, usually on the upper half of the stem. There may be two galls on the same stem, and if you examine the seeds at the top of the stem, you can see that the galls do practically no damage to the plant's ability to reproduce.

Other goldenrod galls: left, *elliptical goldenrod gall;* right, *goldenrod bunch gall*

There are several other common insect galls also seen on winter stems of goldenrod, primarily the elliptical goldenrod gall and the goldenrod bunch gall. The former is an almond-shaped swelling of the stem and the latter is a woody bunch of material at the tip of the stem. (For further

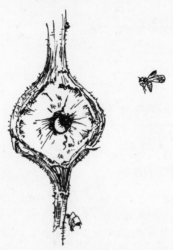

Split-open goldenrod ball gall showing fly larva. On right is adult fly.

information on these, see A *Guide to Nature in Winter*, by this author.)

THE LARVA AND ITS HABITS

If you open the gall, you will see the larva in the center. It is white with a dark head, and by the time winter comes it is fully mature. In spring the larva will become active and chew out a tunnel to the very edge of the gall. It will then go back to the center of the gall and pupate. Flies pupate in their last larval skin, which gets hard and often turns black. This is called a puparium. These flies, along with many of their relatives, have an ingenious way of emerging from the puparium. They have a sac in the front of their heads that is inflated with liquid to press a circular hole through the puparium. The sac is then deflated and retracted. If you look at these galls later in spring, you may find the adult has already emerged and left behind the puparium with its small hole. At this time of year you may also see the exit tunnel created by the larva before it pupates.

BRACONID WASPS

Relationships

Braconid wasps are a family (Braconidae) of insects in the order Hymenoptera, or bees, wasps, and ants. They are extremely small wasps, sometimes no more than an eighth

Habitat of braconid wasp

The "log-pile" configuration of braconid wasp cocoons. Life size

of an inch long. They are closely related to ichneumon wasps, and the two families make up the majority of parasitic insects. Like ichneumons, they tend to lay their eggs on or in other insects, inside of which the young develop and from which they emerge when they are ready to pupate. Many species make tiny white cocoons when pupating and these are at times conspicuous.

Life Cycle

There are thousands of species of braconids, so any account of their life cycle is only a generalization. Most braconid wasps overwinter as larvae or cocoons. The former are usually inside the host and the latter are attached to the outside of the host or nearby on a twig or plant stem. In spring, development resumes with larvae continuing feeding on their host and then pupating in small white or tan cocoons. Adults emerge from the cocoons and mate, and the fertilized females seek out hosts and lay their eggs on or in them. The eggs hatch and the larvae feed on the host. There may be one to three or possibly more generations in a year, the larvae or pupae of the last generation overwintering.

Highlights of the Life Cycle

Practically the only part of the braconid life cycle that is ever seen is the pupal stage, for it is here that the groups of little white cocoons are conspicuous enough to be noticed. Even so, they are small and will only be found by the more industrious insect hunter.

How to Find Braconid Wasp Cocoons

Along the borders of fields or roads, examine the twigs of shrubs and look for clusters of white cocoons, each cocoon no more than an eighth of an inch long. There may be twenty or more cocoons in a cluster and they may be grouped in a log-pile fashion or in an irregular cluster. Sometimes a single cocoon will be found suspended about a half inch below a branch on a single strand of silk. The cocoons may also be found on the cocoons of other insects, such as those of tent caterpillars. Braconid cocoons are probably the hardest to find of all the insects mentioned in this guide.

What You Can Observe

THE COCOONS

To appreciate braconid wasp cocoons, it is best to know the details of the wasp's behavior. Adult braconids feed on nectar from flowers and on some of the honeydew secretions of aphids or treehoppers. After mating, the female, by some as yet unknown mechanism, locates and identifies her host, which in most cases is the caterpillar of a moth or butterfly. Remarkably, each species of braconid wasp usually has a particular species of moth or butterfly that it parasitizes. Once the female finds it, she inserts her ovipositor just under the skin of the host and lays one or more

*Tomato hornworm with braconid wasp cocoons attached to its
outside. Some adult wasps have emerged*

eggs. Sometimes egg-laying is extremely fast, and in one
case a female was recorded as inserting over fifty eggs into
her host in less than a second. In other cases, egg-laying
may take longer.

The larvae hatch inside of the host and begin to feed on
the host's body fluids, and in some cases they do so without
disturbing the vital organs of the host. This enables the
host to continue to live and feed on its own, even with the
parasite inside it. The host insect is essentially eating to
feed both itself and the wasp larvae. By the time the wasp
larvae are ready to pupate, they usually have killed their
host.

Braconid wasp larvae vary in where they form cocoons
and pupate. Some pupate within the host, as is the case
with a common braconid that parasitizes aphids (see
Aphids, summer section). Some attach themselves to the
outside of the host, such as the braconid that parasitizes
the sphinx moth caterpillar (commonly called the tomato
hornworm), frequently found in vegetable gardens in late
summer. Others emerge from their host and pupate singly
or in groups. These are the ones you find in winter. Some
may be found as single cocoons suspended from branches
by a strand of silk. Obviously, the easiest species to spot in
winter are those in which the larvae emerge from the host

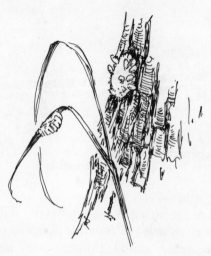

Other arrangements of braconid cocoons: left, *irregular cluster on grass blade;* right, *cocoons attached to tent caterpillar cocoon*

and coordinate the building of their cocoons so that they end up in a log-pile arrangement or random cluster.

This log-pile arrangement of cocoons is one of the most common for braconid wasps. These miniature log piles are usually fastened to a twig. Presumably the host they emerged from was nearby, but I have never found any evidence of a host in winter. The random-cluster configuration of cocoons is also common. These clusters are often found attached to grass stems or sometimes outside the cocoon of the host. Check the cocoons of tent caterpillars for clusters of braconid wasp cocoons arranged on their outer edge.

Some of the cocoons you find may have small holes at one of their ends. This means that the adult wasps have already emerged. On other occasions, some of the cocoons may seem to have been removed from the group, and my suspicion is that they have been pecked away by birds, such as chickadees or titmice.

SPIDER EGG CASES

Relationships

Spiders are not insects but are included here because the egg cases of certain species are common in winter and are usually mistaken for insect cocoons. Spiders are somewhat related to insects in that both are in the phylum Arthropoda. Animals in this phylum all have legs with many

Habitat of garden spiders

Egg-case of orange garden spider. Life size

segments, and external skeletons that are shed each time the animal gets larger. It is easy to distinguish spiders from insects because spiders have eight legs and only two main body parts, whereas insects have six legs and three main body parts.

Life Cycle

Two types of spiders are mentioned here: the garden spiders (genus *Argiope*) and the grass spiders (genus *Agelenopsis*). Garden spiders overwinter as young inside a tough spherical egg sac made by the female in fall. In spring they emerge from the sac and individually make webs and feed as they develop over the summer. By late summer they have matured and mated. The females then make spherical egg sacs and place their eggs in them. Each egg sac is about one inch in diameter and is attached to grass or weed stems. The young hatch in fall but remain in the sac through winter.

Grass spiders, or funnel web spiders, overwinter as eggs placed together in a small convex sac, which is often built in a protected place, such as under loose bark. In spring the eggs hatch and the young build individual webs in which they feed and develop through the summer. By late summer they are mature, and mate. Females lay eggs and construct the covering over them, and the spiders overwinter in the egg stage.

Highlights of the Life Cycle

These spiders can easily be observed throughout the year and are fascinating to study. They are mentioned in this winter section because their egg sacs are often found and puzzled over by people looking for winter insect evidence.

How to Find Spider Egg Sacs

Look for the egg sacs of grass spiders under loose bark of trees. They are a convex mass of white silk about an inch or less in diameter. The silk is sticky and stretches when pulled.

Look for the egg sacs of garden spiders among tall weeds and grasses in lush meadow areas. The sac is a tan sphere, slightly smaller than a Ping-Pong ball, and attached with a few silk threads to the tops of weeds or grasses.

What You Can Observe

GARDEN SPIDER EGG SACS

These egg sacs are most often confused with the cocoons of certain silk moths, for both are large, tan, parchmentlike and found in winter. The spider egg sac was constructed in late fall by the adult female to protect her eggs and young through the winter. The eggs hatch in fall, but the baby spiders remain inside the sac. The stronger ones eat the weaker ones and survive the winter. In spring they make an opening, leave the sac, and go off to make webs and catch prey.

Garden spiders are common and well known. They are large spiders seen in later summer and fall in gardens and weedy fields. Beautifully colored with bands of yellow, black, and white, they make large, circular webs that have a characteristic zigzag pattern across their center. There are two common species and each makes a slightly different egg sac. The orange garden spider (*Argiope riparia*) makes

Garden spider on web

Grass spider egg case under bark

its cocoon basically round with a slight point at one side. The banded garden spider (*Argiope transversa*) makes its egg cocoon round but with one side distinctly flattened. These differences are obvious in the field.

GRASS SPIDER EGG SACS

The other spider construction, found under bark or stones, is a covering for eggs. The silk used for these coverings is of very different quality than that used by insects. As you start to pull at it gently, it is sticky and stretches, and pulls off in tufts like cotton candy. In my experience, insect silk is always more brittle and breaks off. Inside the covering of silk may be a tough little sac, which is a waterproof container for the eggs. As you pull this apart, the eggs may roll loosely onto your hands. They were laid in a more cohesive material, but as they mature they become separate and free inside their container. These egg sacs are made by many species of grass spiders.

The eggs in the papery cocoons of the garden spiders are known to be parasitized by several ichneumon wasps, and these wasps in turn are parasitized by other wasps. It is said that by taking the cocoons inside and keeping them in a closed bottle, you can see the spiders and parasites hatch. The drawback to this is that the young spiders may die, and this would rob you of seeing the spiders and their webs the next summer.

Glossary

Bibliography

Glossary

Adult: An insect that is sexually mature and can reproduce. Also, the final stage in both gradual and complete metamorphosis.

Cocoon: A tough covering made by a larva to protect the pupal phase.

Complete metamorphosis: A type of development that has four distinct stages: egg, larva, pupa, adult.

Generation: The completed life cycle of an insect, from egg to adult.

Gradual metamorphosis: A type of development that has three stages: egg, nymph, adult.

Larva: The second stage in complete metamorphosis, between the egg and pupal stages. Also, the immature insect in that stage.

Molt: An insect's shedding of its skin.

Nymph: The intermediate stage, between egg and adult, in gradual metamorphosis.

Pupa: The generally immobile stage of an insect that is transforming from a larva to an adult.

Bibliography

I once asked a naturalist friend of mine what books I should get to help me learn about insects. She answered, "Any book you can get your hands on." She was absolutely right. There is no one book or even group of books that will give you a complete picture of the lives of insects you commonly find. Below, you will see how varied the sources were for this guide. They span a hundred years and range from popular accounts to detailed scientific studies. Many of the best books are out of print, but can usually be found in stores that sell used books.

Alexander, R. D. 1968. Arthropods. In *Animal Communication*, ed. T. A. Sebeok. Bloomington: Indiana University Press.

Andre, F. 1934. Notes on the biology of *Oncopeltus fasciatus*. *Iowa State College J. Sci.* 9: 73–87.

Balduf, W. V. 1937. Bionomic notes on the common bagworm and its insect enemies. *Ent. Soc. Wash.* 39: 169–84.

——— . 1935. *The Bionomics of Entomophagous Coleoptera*. New York: John Swift Co.

——— . 1939. Food habits of *Phymata pennsylvanica americana*. *Can. Ent.* 71: 66–74.

——— . 1941. Life history of *Phymata pennsylvanica americana*. *Ann. Ent. Soc. Amer.* 34: 204–14.

Bare, C. O. 1926. Life histories of some Kansas "backswimmers." *Ann. Ent. Soc. Amer.* 19: 93–9.

Barrows, E. M. 1976. Mating behavior in Halictine bees: 1, patrolling and age-specific behavior in males. *Jour. Kansas Entomol.* 49: 105–19.

Bilsing, S. W. 1916. Life history of the pecan twig girdler. *Jour. Econ. Entomol.* 9: 110–13.

Borror, D. J., and R. E. White. 1970. *A Field Guide to the Insects*. Boston: Houghton Mifflin.

Borror, D. J., and D. M. DeLong. 1954. *An Introduction to the Study of Insects*. New York: Rinehart and Co.

Brian, M. V., J. Hibble and D. J. Stradling. 1965. Ant pattern and density in a southern English heath. *J. Anim. Ecol.* 34: 545–55.

Campanella, P. J., and L. L. Wolf. 1974. Temporal leks as a mating system in a temperate zone dragonfly. 1: *Plathemis lydia. Behavior* 51: 49–87.

Caponigro, M. A., and C. H. Erikson. 1976. Surface film locomotion by the water strider, *Gerris remigis*. *Am. Mid. Nat.* 95: 268–78.

Caspary, V. G., and A. E. R. Downe. 1971. Swarming and mating of *Chironomus riparius*. *Can. Ent.* 103: 444–7.

Claassen, P. W. 1921. Typha insects: their ecological relationships. *Cornell Univ. Agr. Exp. Sta. Mem.* 47.

Colwell, A. E., and H. H. Shorey. 1975. The courtship behavior of the house fly, *Musca domestica*. *Ann. Ent. Soc. Amer.* 68: 152–6.

Comstock, J. H. 1940. *An Introduction to Entomology*. Ithaca, N.Y.: Cornell University Press.

De La Torre Bueno, J. R. 1917. Life history and habits of the larger waterstrider, *Gerris remigis*. *Ent. News* 28: 201–8.

— — — . 1917. Life history and habits of the margined waterstrider, *Gerris marginatus*. *Ent. News* 28: 295–301.

Dennis, C. J. 1964. Observations on treehopper behavior. *Am. Mid. Nat.* 71: 452–0.

Dew, H. E., and C. D. Michener. 1978. Foraging flights of two species of Polistes wasps. *J. Kansas Ent. Soc.* 51: 380–5.

Dickerson, M. C. 1901. *Moths and Butterflies*. Boston: Ginn & Co.

Dillon, E. S., and L. S. Dillon. 1972. *A Manual of Common Beetles of Eastern North America*. New York: Dover.

Dingle, H. 1968. Life history and population consequences of density, photoperiod, and temperature in a migrant insect, the milkweed bug *Oncopeltus*. *Am. Nat.* 102: 149–63.

Dixon, A. F. G. 1977. Aphid ecology: life cycles, polymorphism, and population regulation. *Ann. Rev. Ecol. Syst.* 8: 329–53.

Downe, A. E. R., and V. G. Caspary. 1973. The swarming behavior of *Chironomus riparius* in the laboratory. *Can. Ent.* 105: 165–71.

— — — . 1973. Some factors influencing insemination in laboratory swarms of *Chironomus riparius*. *Can. Ent.* 105: 291–8.

Downes, J. A. 1969. The swarming and mating flight of Diptera. *A. Rev. Ent.* 14: 271–98.

Eberhard, M. J. W. 1969. The social biology of Polistine wasps. *Misc. Publ. Mus. of Zool. Univ. of Mich.* 20: 755–812.

Emerton, J. H. 1961. *The Common Spiders of the United States*. New York: Dover.

Essenburg, C. 1915. The habits and natural history of the backswimmers *Notonectidae*. *J. Animal Behav.* (Cambridge) 5: 381–90.

Evans, H. E. 1963. *Wasp Farm*. New York: Natural History Press.

— — — . 1966. *Life on a Little-Known Planet*. New York: Dell.

— — — , and M. J. W. Eberhard. 1970. *The Wasps*. Ann Arbor: University of Michigan Press.

Felt, E. P. 1926. *Manual of Tree and Shrub Insects*. New York: Macmillan Co.

— — — . 1940. *Plant Galls and Gall Makers*. Ithaca, N.Y.: Comstock.

Fitzgerald, T. D. 1976. Trail marking by larvae of the eastern tent caterpillar. *Science* 194: 961–3.

Frost, S. W. 1959. *Insect Life and Insect Natural History*. New York: Dover.

Funkhouser, W. D. 1917. Biology of the Membracidae of the Cayuga lake basin. *Cornell Univ. Agr. Exp. Sta. Mem.* 11: 177–445.

Gibson, W. H. 1892. *Sharp Eyes*. New York: Harper & Brothers.
——— . 1897. *Eye Spy*. New York: Harper & Brothers.
Gittelman, S. H. 1974. Locomotion and predatory strategy in backswimmers. *Am. Mid. Nat.* 92: 496–500.
Hagen, K. S. 1962. Biology and ecology of predaceous Coccinellidae. *Ann. Rev. Ent.* 7: 289–326.
Hayes, W. P. 1919. The life cycle of *Lachnosterna lanceolata*. *Jour. Econ. Entomol.* 12: 109–17.
Heinrich, B. 1975. Energetics of pollination. *Ann. Rev. Ecol. Syst.* 6: 139–70.
——— , and F. D. Vogt. 1980. Aggregation and foraging behavior of whirligig beetles. *Behav. Ecol. Sociobiol.* 7: 179–86.
Hodek, I. 1966. *Ecology of Aphidophagous Insects*. Prague: Academia.
Holland, W. J. 1903. *The Moth Book*. Garden City, N.Y.: Doubleday, Page and Co.
Howard, L. O. 1914. *The Insect Book*. Garden City, N.Y.: Doubleday, Page and Co.
Jacobs, M. E. 1955. Studies on territorialism and sexual selection in dragonflies. *Ecology* 36: 566–86.
Kerr, G. E. 1974. Visual and acoustical communicative behavior in *Dissoteira carolina*. *Can. Ent.* 106: 263–72.
Klots, E. B. 1966. *The New Fieldbook of Freshwater Life*. New York: G. P. Putnam's Sons.
Lanham, U. 1964. *The Insects*. New York: Columbia University Press.
Larson, P. P., and M. W. Larson. 1965. *All About Ants*. New York: Thomas Y. Crowell Co.
Lavigne, R. J., and D. S. Dennis. 1975. Ethology of *Efferia frewingi* (Diptera: Asilidae). *Ann. Ent. Soc. Amer.* 68: 992–6.
Linsley, E. G. 1940. Notes on Oncideres twig girdlers. *Jour. Econ. Entomol.* 33: 561–3.
Lloyd, J. E. 1966. Studies on the flash communication system in Photinus fireflies. *Misc. Publ. Mus. Zool., Univ. Mich.* 130: 1–95.
Lyford, W. H. 1975. Overland migration of Collembola colonies. *Am. Mid. Nat.* 94: 205–9.
Maier, C. T., and G. P. Waldbauer. 1979. Dual mate-seeking strategies in male syrphid flies. *Ann. Ent. Soc. Am.* 72: 54–61.
Malyshev, S. I. 1935. The nesting habits of solitary bees. *Eos* (Madrid) 11: 201–307.
Matheny, W. A. 1909. The twig girdler. *Ohio Nat.* 10: 1–7.
McColloch, J. W., W. P. Hayes, and H. R. Bryson. 1928. Hibernation of certain scarabaeids and their Tiphia parasites. *Ecology* 9: 34–42.
Metcalf, C. L. 1911. Preliminary report on the life histories of two species of Syrphidae. *Ohio Nat.* 11: 337–46.
——— . 1912. Life histories of Syrphidae III. *Ohio Nat.* 12: 477–89.
——— , and W. P. Flint. 1928. *Destructive and Useful Insects*. New York: McGraw-Hill.
Milne, L., and M. Milne. 1980. *The Audubon Society Field Guide to North American Insects and Spiders*. New York: Alfred A. Knopf.
——— . 1980. *Insect Worlds*. New York: Charles Scribner's Sons.

Mitchell, R. T., and H. S. Zim. 1962. *Butterflies and Moths.* New York: Golden Press.

Morris, R. F. 1972. Predation by insects and spiders inhabiting colonial webs of *Hyphantria cunea. Can. Ent.* 104: 1197–1207.

——— . 1972. Predation by wasps, birds, and mammals on Hyphantria cunea. *Can. Ent.* 104: 1581–91.

Myers, J. H., and J. N. M. Smith. 1978. Head flicking by tent caterpillars: a defensive response to parasite sounds. *Can. J. Zool.* 56: 1628–31.

Nuhn, T. P., and C. G. Wroght. 1979. An ecological survey of ants in a landscaped suburban habitat. *Am. Mid. Nat.* 102: 353–62.

Oldroyd, H. 1964. *The Natural History of Flies.* New York: W. W. Norton & Co.

Packard, A. S. 1873. *Our Common Insects.* Boston: Estes and Lauriat.

Palmer, E. L. 1949. *Fieldbook of Natural History.* New York: McGraw-Hill.

Pliske, T. E. 1975. Courtship behavior of the monarch butterfly. *Ann. Ent. Soc. Am.* 68: 145–51.

Richards, K. W. 1978. Nest site selection by bumblebees in southern Alberta. *Can. Ent.* 110: 301–18.

Rose, A. H., and O. H. Lindquist. 1973. Insects of eastern pines. Dept. of the Environment, Can. For. Ser. Publ. 1313.

Rudinsky, J. A. 1962. Ecology of Scolytidae. *Ann. Rev. Entomol.* 7: 327–48.

Rudinsky, J. A., P. T. Oester, and L. C. Ryker. 1978. Gallery initiation and male stridulation of the polygamous spruce beetle *Polygraphus rufipennis. Ann. Ent. Soc. Am.* 71: 317–21.

Rutowski, R. L. 1978. The form and function of ascending flights in *Colias* butterflies. *Behav. Ecol. Sociobiol.* 3: 163–72.

Scarbrough, A. G. 1978. Ethology of *Cerotainia albipilosa* (Diptera: Asilidae) in Maryland: Predatory behavior. *Proc. Entomol. Soc. Wash.* 80: 113–27.

Scarbrough, A. G., and A. Norden. 1977. Ethology of *Cerotainia albipilosa* (Diptera: Asilidae) in Maryland: diurnal activity rhythm and seasonal distribution. *Proc. Entomol. Soc. Wash.* 79: 538–54.

Scarbrough, A. G., J. G. Sternburg, and G. P. Waldbauer. 1977. Selection of the cocoon spinning site by the larvae of *Hylaphora cecropia* (Saturnidae). *J. Lepid. Soc.* 31: 153–66.

Schmitz, R. F. 1972. Behavior of Ips pini during mating, oviposition, and larval development (Coleoptera: Scolytidae). *Can. Ent.* 104: 1723–8.

Scudder, S. H. 1899. *Everyday Butterflies.* Boston: Houghton Mifflin.

Shelford, V. E. 1908. Life histories of the tiger beetles. *Journ. Linn. Soc. Zool.* 30: 157–81.

Sheppard, R. F., and G. R. Stairs. 1976. Factors affecting the survival of larval and pupal stages of the bagworm, *Thyridopteryx ephemeraeformis. Can. Ent.* 108: 469–73.

Smith, R. C. 1921. A study of the biology of the Chrysopidae. *Ann. Ent. Soc. Am.* 14: 27–35.

Snodgrass, R. E. 1930. *Insects: Their ways and means of living.* In Smithsonian Scientific Series, ed. C. G. Abbot. New York: Smithsonian Institution Series, Inc.

Solomon, J. D. 1977. Frass characteristics for identifying insect borers in living hardwoods. *Can. Ent.* 109: 295–303.

Solomon, J. D., and W. K. Randall. 1978. Biology and damage of the willow shoot sawfly in willow and cottonwood. *Ann. Ent. Soc. Am.* 71: 654–7.

Spence, J. R., D. H. Spence, and G. G. E. Scudder. 1980. Submergence behavior in Gerris: underwater basking. *Am. Mid. Nat.* 103: 385–91.

Stokes, D. W. 1976. *A Guide to Nature in Winter.* Boston: Little, Brown.

Swain, R. B. 1948. *The Insect Guide.* Garden City, N.Y.: Doubleday.

Swan, L. A., and C. S. Papp. 1972. *The Common Insects of North America.* New York: Harper and Row.

Teskey, H. J. 1969. On the behavior and ecology of the face fly, *Musca autumnalis. Can. Ent.* 101: 561–76.

Tilman, D. 1978. Cherries, ants, and tent caterpillars: timing of nectar production in relation to susceptibility of caterpillars to ant predation. *Ecology* 59: 686–92.

Travis, B. V. 1939. Habits of the june beetle, *Phyllophaga lanceolata,* in Iowa. *Journ. Econ. Entomol.* 32: 690–7.

Urbani, C. B. 1979. Territoriality in social insects. In *Social Insects,* ed. H. R. Hermann. New York: Academic Press Inc.

Urquhart, F. A., and N. R. Urquhart. 1976. The overwintering site of the eastern population of the monarch butterfly in southern Mexico. *J. Lep. Soc.* 30: 153–8.

— — — . 1978. Autumnal migration routes of the eastern population of the monarch butterfly. *Can. J. Zool.* 56: 1759–64.

Waage, J. K. 1973. Reproductive behavior and its relation to territoriality in *Calopteryx maculata* (Odonata: Calopterygidae). *Behavior* 47: 240–55.

Wallis, D. I. 1961. Behavior patterns of the ant, *Formica fusca. Anim. Behav.* 10: 105–11.

— — — . 1961. Aggressive behavior in the ant, *Formica fusca. Anim. Behav.* 10: 267–74.

Way, M. J. 1963. Mutualism between ants and honeydew-producing Homoptera. *Ann. Rev. Entomol.* 8: 307–44.

Weed, C. M. 1897. *Life Histories of American Insects.* New York: MacMillan Co.

— — — . 1901. *Nature Biographies.* Garden City, N.Y.: Doubleday, Page and Co.

— — — . 1917. *Butterflies Worth Knowing.* Garden City, N.Y.: Doubleday, Page and Co.

Wellington, W. G. 1974. Changes in mosquito flight associated with natural changes in polarized light. *Can. Ent.* 106: 941–8.

Wheeler, E. W. 1923. Some braconids parasitic on aphids and their life history. *Ann. Ent. Soc. Am.* 16: 1–28.

Wilcox, R. S. 1972. Communication by surface waves: mating behavior of a water strider. *J. Comp. Physiol.* 80: 255–66.

Wilson, E. O. 1971. *The insect societies.* Cambridge, Mass.: Harvard University Press.

Wood, T. K., and R. L. Patton. 1971. Egg froth distribution and deposition by *Enchenopa binotata* (Homoptera: Membracidae). *Ann. Ent. Soc. Am.* 64: 1190–1.

Zorn, L. P., and A. D. Carlson. 1978. Effect of mating on response of female *Photuris* firefly. *Anim. Behav.* 26: 843–7.

Insects according to orders:

DRAGONFLIES AND DAMSELFLIES
White-tailed dragonfly
Black-winged damselfly

GRASSHOPPERS AND CRICKETS
Crickets and grasshoppers
Tree crickets
Mantids

BUGS
Water striders
Large milkweed bug
Backswimmers
Ambush bugs

APHIDS, CICADAS, ETC.
Spittlebugs
Cicadas
Leafhoppers
Treehoppers
Aphids
Woolly alder aphids

BEETLES
Whirligig beetles
June beetles
Willow and cottonwood leaf beetles
Tiger beetles
Fireflies
Tortoise beetles
Longhorned and soldier beetles
Acorn weevils
Twig girdlers and twig pruners
Ribbed pine borer
Bark beetles